AF470933

Science and hill farming

Twenty-five years of work at the
Hill Farming Research Organisation
1954-1979

Published by the Hill Farming Research Organisation on
the occasion of their Silver Jubilee, September 1979

ISBN 0 7084 0117 1

Board of Management
as at 1 April 1979

Chairman	Mr R David Ker
	Mr C H Armstrong
	Professor D G Armstrong MSc PhD DSc FRIC FIBiol FRSE
	Mr A Evans
	Mr J W Grant BSc(Agr) NDD
	Professor J S Hall CBE BSc(Agr) FRAgS FIBiol
	Mr L M Howells
	Mr J Manuel
	Mr N L McCall Smith JP
	Mr A S Macdonald
	Dr D R Melrose BSc DVMS FRCVS
	Mr R L Ollerenshaw OBE JP
	Mr J A Parry MA Vet MB MRCVS
	Professor J H D Prescott BSc(Agr) PhD
	Professor M B Wilkins PhD DSc FRSE

Staff
as at 1 April 1979

Director
J M M Cunningham CBE BSc(Agr) PhD FIBiol FRSE

Animal Production and Nutrition Department
Head of Department J Eadie BSc(Agr)

R D M Agnew HNC LIBiol
Rchd H Armstrong BSc(Agr)
M M Beattie HNC
Mrs S F Campbell
Miss E V Deans
J M Doney BSc PhD
A R Fawcett AIMLS
Mrs H Fisher ONC
R G Gunn BSc(Agr) PhD
M Hisbent ONC
J Hodgson BSc(Agr) PhD
C D Kerr NDA
C S Lamb BSc(Agr)
I D Leslie
A J Macdonald NDA
D N McFarlane BSc(Agr)
T J Maxwell BSc PhD
R W Mayes BSc MSc PhD

J A Milne BSc(Agr) PhD
Miss P E Moberly BSc
J N Peart BSc(Agr) MSc
I R Pitkethly HND
S J Rhind BSc PhD
J Riva
A J F Russel BSc M Agr Sc PhD
Mrs C Scott
A R Sibbald HNC
D A Sim ONC
E Skedd HNC
A D M Smith SDA NDA
W F Smith AIST
Mrs A M Spence BA
A P Thompson BSc(Agr Biochem)
I R White HNC
A Whitelaw BSc FRCVS
S Wilson BSc(Agr) MS

Plants and Soils Department
Head of Department P Newbould BSc BAgr DPhil

G J Baillie
R Begbie ONC
A M Bryce
A R M Chambers BSc
G E Davies BSc
C C Evans HNC
M J S Floate BSc PhD ARIC
Miss S A Grant BSc MSc
A Haystead BSc PhD
A D Ironside ONC
J King BSc(Agr) PhD
W I C Lamb HNC MIBiol

Mrs J Leask
J Mackenzie HNC
Mrs C A Marriott BSc
Miss A Rangeley BSc
J A Rogers BSc PhD
Miss S Sharrock
Miss E M Smith
D E Suckling HNC MIBiol
Mrs C Thompson
Mrs M Thornton
Miss E Tierney
Miss L Torvell BSc

Librarian/Information Officer Mrs M C Pitkethly BSc MSc

Administration
Secretary H C M McLeod

Miss I C R Anthony
Miss J A Brown
Miss J C Brown
Mrs E M Campbell
Miss J P Hall
Mrs E B H Reid

Mrs J Scotland
Mrs H Tulloch
F Ward M Inst P&S
Mrs M J Williamson
Mrs S M G Young

Maintenance
Caretaker H Thompson

W Adamson
A Black
Mrs A Cameron
Mrs A Conway
Mrs A Davies

Mrs D Fairley
Mrs A McManus
Mrs A O'Hara
Mrs M Page
Mrs A Young

Glensaugh Research Station
Officer-in-charge W J Hamilton BA NDA NDD MIBiol

Scientific staff
Mrs E Barthram BSc(Agr)
G T Barthram BSc
M Begg OND

A J Senior SDA
W G Souter ONC

Farm staff
K Adams
Mrs K B Cargill
Miss L Fairlie
H Habblett
D Harrison
R Jack

N McEwan
A R Smith
Mrs B Smith
Miss L A Stokes
C Thomson

House o'Muir Research Station and Animal House, Bush
Officer-in-charge A L Fairley BSc(Agr)

A W Hetherington
J Leary
C McCormack

N W Mortimer
J Smith
R Smith

Lephinmore Research Station
Officer-in-charge D C Currie NDA NDD

Scientific staff
G R Bolton BSc(Agr)

T K Whyte HNC SDA

Farm staff
J W Black
Mrs J P Black
B Macfarlane

G Scott
Mrs K A Scott

Sourhope Research Station
Officer-in-charge R H Armstrong BSc MSc PhD ARIC

Scientific staff
D R Campbell SDA NDA
T G Common HNC
M P Gill BSc

R A Hetherington BSc(Agr)
A McFadzen BSc

Farm staff
E McFarlane
P Marshall
A Ramage
Miss E M Smales
N Straughan
M Taylor

A G Todd
Mrs A S Todd
S Townsend
Mrs J Treasure
J Wallace

ARC Unit of Statistics

G J Davies BA

Post-graduate Students

J Bircham M Agr Sc
T D A Forbes BA MSc
Mrs L J Sheppard BSc
I A Wright BSc

We also acknowledge the contribution of previous board members and staff, too numerous to mention individually, who have served the Organisation during the past 25 years.

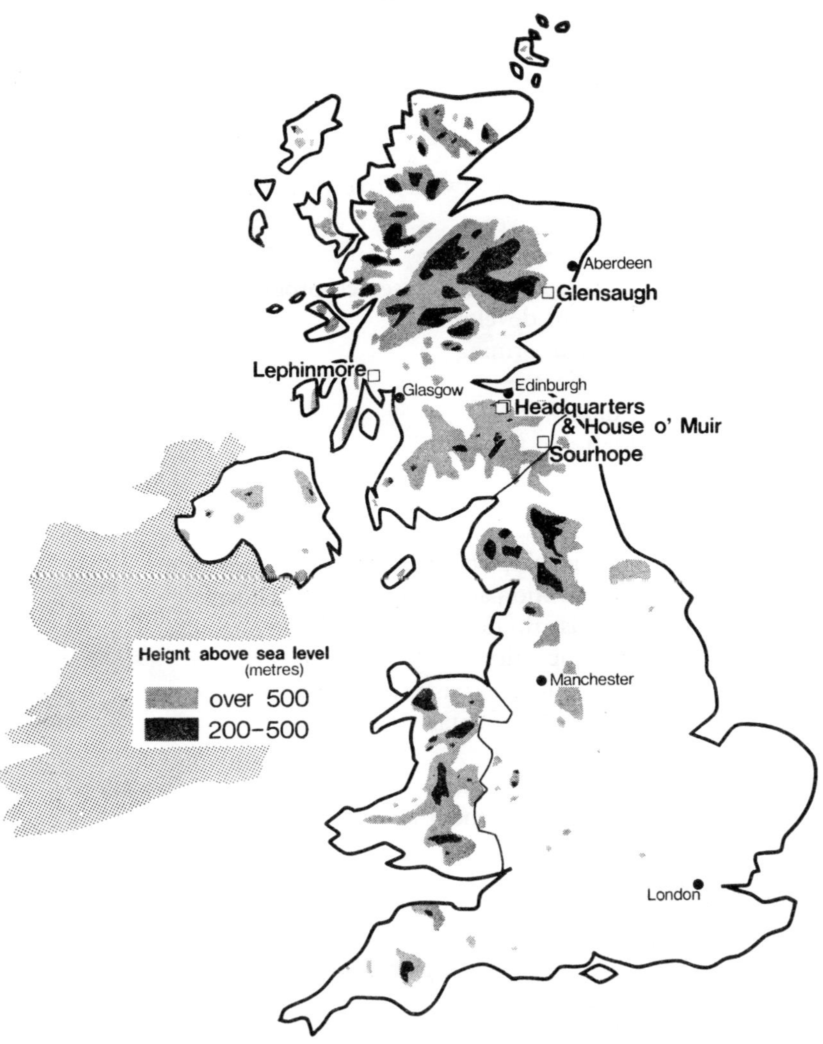

Location of HFRO research stations

Contents

Preface

The Board of Management decided that it would be appropriate to mark, in a suitable manner, the Silver Jubilee of the Hill Farming Research Organisation. It was proposed that a review of the work of the Organisation over the past 25 years should be published.

When brought together in this book the depth and width of the scientific effort makes interesting and impressive reading and is of practical significance for the future development of the hill farming industry. Although work has been concentrated in Scotland, the aim of research has been to obtain an understanding of the factors influencing plant and animal production which would be applicable to the hill areas in the United Kingdom. Indeed, the principles derived from research have a wide application and are equally relevant to sheep and cattle production in many other areas of the world, where much of the scientific work has received international recognition. The truth of this is confirmed by the repeated requests for members of HFRO staff to visit and convey their knowledge to all corners of the world. During a recent visit to the Falkland Islands — some 8000 miles distant — I was most interested to see research work in the Grassland Trials Unit being conducted under direction from HFRO in response to a request some years ago by the Ministry of Overseas Development. Interest in the Organisation is also evident from the large number and constant flow of visitors received from all over the world.

As one moves around the countryside of Scotland, England and Wales, it is most encouraging to see the vast improvement which has taken place in many hill areas — whole hillsides converted from poor, unproductive pasture into good grazing, fenced and carrying a greatly increased stock of both sheep and cattle. I believe that much of this improvement has been stimulated by the research of HFRO being transferred into the field through the advisory services. There is no doubt whatsoever that over the last 25 years stock on the hills, both sheep and cattle, have improved vastly both in condition and health and, consequently, in production. The Board of Management believe that HFRO has made a significant impact on hill farming but vast areas of the United Kingdom are still crying out for improvement and development and there is a great potential for the future.

The Hill Farming Research Organisation was conceived in times of extreme financial difficulties in hill and upland farming. Development of the industry will require capital to take advantage of the results of research work.

The former Directors, the late Mr Arthur Wannop, OBE, and Professor R L Reid, both brought leadership, direction and sound guidance. Mr Wannop, in the formative years, laid a sound foundation including an understanding of the traditional systems, while Professor Reid was responsible for initiating the programmes of more critical investigations central to the scientific advances which have been achieved.

Wise advice and encouragement from the previous Chairmen of the Board of Management, Lord Stratheden and Campbell and Mr J Crozier, and the unfailing interest and support of Board members have contributed greatly to the enthusiasm and commitment of staff.

The present Board wish to express their appreciation of all the work which has been done. This appreciation extends, most sincerely, to the whole staff of the Organisation at headquarters and on the research stations who carry on their work efficiently and with real interest, led by the present Director, Professor J M M Cunningham.

The Board would also wish to place on record their appreciation of the support of the Department of Agriculture and Fisheries for Scotland in providing funds and facilities to allow the development of a sound research commitment, and also to the Agricultural Research Council for advice in the planning of research strategy and in support of projects.

We present this book as a review of the work of the Organisation over the past 25 years. Understandably it does not refer in detail to, or necessarily quote from, all of the research activities which have taken place during that time, but rather attempts to present a logical picture of the way in which our understanding of hill and upland agriculture has developed. We recognise that much valuable work has been done elsewhere, but reference to this has been kept to a minimum as the aim is to present an overall picture of the Organisation's contribution to the science of hill farming, rather than a comprehensive review of the subject.

Much of the research has been undertaken in clearly defined areas and the chapters in this book reflect this. Several staff have contributed to each chapter, the preparation of which has been the responsibility of a co-ordinator. The complete publication has been planned and edited by a panel comprising: Miss S A Grant (Convener), Dr J M Doney, Dr A Haystead, Mrs M C Pitkethly, Dr A J F Russel, Mr A R Sibbald and Mrs H Tulloch; the text figures and photographs were prepared by Mr I R Pitkethly. To all of these we are extremely grateful for the special effort which they have made.

R David Ker,
Chairman of the Board of Management

Introduction

Hill farming, so important in Scottish agriculture and extending through all the hilly and mountainous areas from the Border to the Northern Isles, was in serious economic straits in the early years of the last war. The concern about its viability, and its future, was so widespread that Tom Johnston, then Secretary of State for Scotland, set up a committee of enquiry under the chairmanship of Lord Balfour of Burleigh to look into the problems facing this branch of farming and to make recommendations. Similar worries about the position of hill farming in England and Wales led the Minister of Agriculture to appoint a similar committee of enquiry.

The Balfour of Burleigh Committee (GB Parl, 1944) included among its many recommendations quite a number dealing with future research. It was in these recommendations that public expression was first given to the desirability of making a centralised and co-ordinated research attack on the scientific, technical, and economic problems facing hill farming.

The problems of hill farming had not by any means been neglected by agricultural research workers or by the agricultural colleges in their experimental and advisory work. Indeed, in earlier years, some of the most spectacular pieces of research had greatly benefited hill farmers. This was particularly true in the field of animal diseases where, for example, preventive measures for lamb dysentry and for braxy and louping-ill had been discovered and put into practice. To the Balfour of Burleigh Committee this was indicative of what might follow a properly co-ordinated research approach to the whole range of problems from which the industry was suffering.

In the forties hill farming was probably the most rigid and intractable form of husbandry that can be imagined. The Committee said in its report 'in no other type of farming is there a stronger belief in traditional methods (in spite of the adoption of some novelties in the treatment and prevention of disease) or such a fatalistic acceptance of misfortune'.

The central recommendation of this Committee on the subject of research was that there should be centralised control of research into the industry's problems and the establishment of a station for co-ordinated research work. At that time, in the middle of a war, with no likelihood of significant new financial resources becoming available for some time to come, the form of control proposed was by a standing Hill Farm Sub-Committee of the Scottish Agricultural Improvement Council. The idea of one research station, which seems ludicrous now, must also have reflected the exigencies of the times. The station was to provide facilities for applied research based on more fundamental work done at universities and research institutes as a prelude to a much greater development of advisory work among hill farmers by the agricultural colleges.

There was no hesitation by the powers-that-be in accepting these recommendations and, in the light of the emphasis which had also been placed on research in the report of the English and Welsh Committee of Enquiry, discussions in the Agricultural Departments and in the Agricultural Research Council went ahead on a Great Britain basis, and not solely in a Scottish context. Naturally, with hill farming so much more important to Scottish agriculture, the initiative in development lay in Scotland.

An *ad hoc* committee of the Scottish Agricultural Improvement Council quickly reported in 1944 on how best to give effect to the Balfour of Burleigh Committee's research recommendations (DOAS, 1951). It endorsed the concept of committee management of research with a specially selected scientific officer who would be the secretary of the management committee and also its link with research institutes, colleges, and the industry. But the Committee also recommended that there should be three research stations, one at Glensaugh, which the North of Scotland College of Agriculture had acquired in 1943, another in the west of Scotland, perhaps Argyll, and a third in the Borders which could serve the needs of the north of England as well as the south of Scotland.

In November 1945 a Hill Farm Research Committee of the Scottish Agricultural Improvement Council was appointed by the Secretary of State for Scotland. It included members from England and Wales and had a balanced representation from universities, institutes and colleges, as well as from the industry. In the period of this Committee's existence wide-ranging reviews of the industry's problems were undertaken and the three research stations established. Sourhope, on the border with England, was leased in 1946, while Lephinmore, a 'remnant' from a forestry acquisition, was made available by the Secretary of State in 1948.

Under management of research by committee the only practicable way of handling the day-to-day running of the research stations was to place them under the control of the agricultural college in whose area they were situated. The North of Scotland College of Agriculture was already managing Glensaugh. This arrangement was not meant to derogate in any way from the Hill Farm Research Committee's responsibility for the co-ordination and oversight of the work on the stations. This Committee reported on its work in 1951, but is was only when it produced its Second Report (DOAS, 1952) that it was made abundantly clear that the full management of the three research stations should be under central control together with control of the research.

Looking back it was clearly inevitable that the stations, while under the control of the colleges, should become increasingly used for experiment and demonstration work to the detriment of anything more fundamental and detrimental also to the universities and research institutes for whom they were to provide field facilities. At the same time it was difficult to carry out work on the stations which ran contrary to the accepted standards of good husbandry, which can often be essential in research. Accordingly, in its Second Report the Hill Farm Research Committee recommended that there should be centralised control and management of the three stations, as well as scientific control by an independent committee. This committee, though not a governing body like those at the research institutes, should, nevertheless, functon as such with the help of a senior scientist of director status.

It is a little difficult now to understand why this last recommendation should have stopped short of proposing a new research institute. But it is clear from the record of discussions at that time that the idea of an independent institute for research into the problems of a type of farming was very difficult for the authorities to accept. With the exception of those concerned with dairying, research institutes had always been established on the basis of a subject or a scientific discipline.

So in 1952, when the recommendations in this Second Report were considered, discussions again began with the idea of setting up another Hill Farm Research Committee on the lines proposed. However, discussions were somewhat prolonged and, fortunately, the 'yeast' of common sense began to work. When action was finally taken in December 1953 the announcement was made of the appointment of a Governing Body for a Hill Farming Research Organisation, to be headquartered in Scotland, but with Great Britain as its field of interest. And so, at last, on 1st April 1954, a new research institute was born — the Hill Farming Research Organisation as we know it today — with Arthur Wannop, OBE, as its first Director.

It may seem a long time from 1944, when the Balfour of Burleigh Committee reported, to 1954 when the Hill Farming Research Organisation was set up. However, this apparently slow progress has to be viewed against the background of the difficulties of the post-war years and the prejudices and weight of tradition that had to be overcome, both in the scientific world and in the industry itself.

W H Senior, CB, MSc, FRSE,
former Under Secretary at the
Department of Agriculture for Scotland.

References

DOAS (1951). *Hill Farm Research.* Report of the Scottish Hill Farm Research Committee. Edinburgh, HMSO.

DOAS (1953). *Hill Farm Research Second Report.* The Scottish Hill Farm Research Committee. Edinburgh, HMSO.

GB PARL (1944). Cmd 6494. DOAS. *Report of the Committee on Hill Sheep Farming in Scotland.* Edinburgh, HMSO.

Main building

Headquarters facilities

Glasshouses

Animal house

1

A perspective of twenty-five years of hill farming research

Contributor: Professor J M M Cunningham

The early years

At its inception in April 1954 the Hill Farming Research Organisation was established as an independent grant-aided institute under the control of a Board of Management. This is made up of representatives of hill farming from Scotland, England and Wales, and approximately equal numbers of scientists whose field of specialisation is relevant to the research programmes in the Organisation.

Under its Articles of Association the Organisation was established 'to conduct and promote research on the problems arising from hill farming . . . including possible methods of increasing agricultural production and of eradicating disease or pests on hill land'. In the planning and conduct of work the research programmes have been subjected to the scientific supervision of the Agricultural Research Council.

It was visualised at the outset that the primary function of the Organisation would be field research requiring large numbers of sheep; the provision of three relatively large hill farms was accordingly arranged. Administrative offices, with only limited laboratory facilities, such as would be necessary to complement the actual research in the field, were established in Edinburgh in 1954.

The initial intention was that HFRO would apply the basic knowledge obtained in the more discipline-orientated research institutes to an applied experimental programme. Essentially the policy appeared to place emphasis on the husbandry-type trial and also on what is nowadays described as development work.

At that time no systematic study of the problems of hill sheep production had been undertaken but accounts of the improvers and leaders in hill farming were to be found in the journals of the agricultural societies and elsewhere.

Over a period of thirty years, after the formation of the Welsh Plant Breeding Station in the mid-twenties and arising out of the work of Sir George Stapledon, a considerable amount of effort was devoted to the problems of hill improvement in Wales. The responses of indigenous hill pasture communities to the application of lime and fertiliser had been measured and the potential of new grasses and clovers assessed. This phase at the Welsh Plant Breeding Station was followed by work which concentrated on studying management effects, including the influence of the grazing animal on the production and maintenance of improved swards.

The higher proportion of relatively good improvable soils, the smaller farms and hence greater economic pressure, and the enthusiastic and persuasive promotion of land improvement by Stapledon with a pioneer zeal, combined to cause more rapid changes in Wales than elsewhere.

Hill sheep farming

Hill farming is not uncommonly used to describe both hill sheep and upland farms. Traditionally hill sheep farms have self-replenishing flocks of the hill breeds kept in regular ages, and in some instances bound to the ground, so that when ownership or tenancy changes the stock remain. These farms are predominantly rough grazings having 95% or more of the land in this category. Ewe flocks are maintained in some parts of the country in 'hefts' or 'cuts' which are sub-flocks habitually grazing within the confines of a particular area of ground. The sheep are dependent on grazed herbage throughout·the year. In the mid-fifties supplementary feeding was uncommon, apart from during spells of complete snow cover, but it has now become more widespread especially during late pregnancy.

Stocking rates are generally low, 0.75-4.0 ha/ewe, being determined by the winter carrying capacity which is set at a level offering a sufficient selection of herbage to avoid excessive winter live-weight loss.

Pastures comprise a wide variation in plant associations, reflecting the underlying soils, the topography and climate, which are described in Chapter 2. There is a wide diversity between farms and within farms, and even within a single hillside.

The traditional grazing management practice varies widely from allowing the sheep to graze where they will, to daily routine 'raking', i.e. the movement of ewes to graze the lower ground during the day and to higher ground in the late afternoon. Variations in winter management occur in

Wales, where the sheep are moved into a paddock or 'ffridd' on the lower hill, and in the north east of Scotland where arable crops, mainly roots, are used as part of the winter feed.

These systems are characterised by low levels of output both on a per head and per hectare basis, cattle play a varying although usually minor role, so that a significant proportion of the production — store or finished lambs, wool and draft ewes — is derived from sheep. Weaning percentages vary from around 60 to 100 or above on the best hill farms. Average reproductive rate is probably about 80% and, with weaning weights of approximately 24 kg per lamb weaned, lamb output of 15 kg/ha is common. Thus, low reproductive performance, poor lamb growth rate, high mortality of lambs and ewes, low stocking rates and hence poor utilisation of pastures, contribute in part or whole to the marginality of hill agriculture. For decades the hill sheep farming industry has been beset by economic problems, and for many years has received Government support in the shape of subsidies and grants. During some periods, costs, particularly of labour, have tended to rise at a relatively faster rate than end-product prices. The average hill farm in Scotland, where they are generally larger than in the remainder of the United Kingdom extends to 1600 hectares, carries 1030 ewes and 27 beef cows. Although apparently large scale, gross output is considerably less than for other farm categories and, while net output is proportionally similar, overall economic business size is small and therefore more vulnerable to economic pressures.

All of these farms fall within the category of Less Favoured Areas, as described in Article 3.4 of the EEC Directive 268 (EEC, 1975) which defines these areas as being on infertile land suitable only for extensive livestock farming, as being in danger of depopulation, and where the conservation of the countryside is necessary.

Development of a research programme

It was understandable therefore that the attention of HFRO was concentrated on problems of animal production as well as those of plant production. Indeed, at the time of the Organisation's inception little basic information existed on many aspects of sheep production, and to establish baselines of animal performance under traditional systems, routine recording was established on all research stations and has continued to a greater or lesser extent to the present time.

Perhaps two of the original areas of investigation illustrate how the priorities and organisation of research programmes have changed over the years.

The Hill Farm Research Committee (DOAS, 1951) recommended a

need for research centred on the nutrition of the hill ewe since she is required to survive during the winter on a level of nutrition which is frequently insufficient to maintain live weight whilst her annual pregnancies coincide with a period of semi-starvation. Field-scale feeding trials were instituted and, while these produced useful practical guidelines, the scientists involved were to comment that it was difficult to assess what constituted an excessive loss in weight. It was concluded that the elucidation of the more fundamental aspects of the problem would be essential, requiring controlled conditions. This knowledge was not being obtained elsewhere and so in the early sixties the Organisation began to create facilities to undertake the more basic studies on the nutrition of the pregnant ewe and, later, on lactation and fertility. The work in these areas is reviewed in Chapter 4. Although the practice of winter feeding is now more widespread as a result of the Organisation's early work, research on the nutrition of the pregnant ewe still continues at both a basic and applied level with a view to obtaining more specific understanding relevant to different types of hill wintering. Meanwhile other studies on lactation, lamb growth and ewe fertility all pointed to an unexploited potential in hill sheep arising from deficiencies in summer and autumn nutrition.

The second example concerns work on hill pastures which was devoted initially to finding ways of improving the utilisation and management of hill vegetation (Chapter 5). The early studies on the grazing preferences of sheep for distinct plant communities led to investigation of nutritional quality on a seasonal basis of the more important ones. Studies on the nutritive value of an increasing range of hill species and communities were undertaken over the years and continue today, including, within recent years, cattle.

Systems development

By the late sixties the accumulated understanding made possible a synthesis of a new hill sheep system, subsequently to become popularly known as the 'two-pasture system'. This immediately led to a substantial change in the research strategy on hill sheep in the Organisation. It created a more specific context for the provision and utilisation of both improved and indigenous pasture. The development of econometric techniques highlighted those aspects of animal performance which are important, thus allowing a clearer definition of research priorities.

Until the limitations to animal production in the hills were understood and the best way of integrating improved pasture into a true hill system was developed, land improvement made only minimal impact on hill sheep production.

Since 1968 the development and testing of complete hill sheep production systems has been an integral part of our research programme. This systems research has been undertaken on a farm scale with a substantial commitment of land and staff on the research stations where the officers-in-charge have had a major responsibility for supervision of the programme. Indeed this work has been supported by many of the scientific, technical and farm staff. Apart from creating the conditions for the scientists to see their work applied in the field, the systems development programme has involved intensive monitoring of animal health and performance. The application, for example, of ideas regarding the control of liver fluke, now routine practice within the development project at Lephinmore, has demonstrated the merits of the proposed practices on an adequate field scale. Interpretation of the data from the development projects permits the identification of problem areas and assists in the planning of the analytical research programme.

Upland farming

In 1972 HFRO embarked upon a new area of work when the remit was extended to include the problems of upland farming.

Contrasted with hill farms, upland farms have a much higher proprotion of enclosed sown pastures, not uncommonly 30% or more, associated with access to rough grazing either as part of the farm or with rights of access to common grazing. The latter make a much smaller contribution to animal production than in hill systems. Conservation of winter fodder is possible on many farms, but the problems of access and slope impose limitations not present to a similar extent on lowland farms. With enclosed pastures at elevations often in excess of 200 m, the period of herbage growth is short, commencing in mid-April to early May and ceasing by the beginning of October. Production is less prone to being affected by drought, indeed some soils give rise to problems of wetness imposing limitations, particularly for winter stocking. Income is derived mainly from breeding stock, both sheep and cattle.

On upland farms sheep are generally crossbred ewes derived from hill breeds (Scottish Blackface, Cheviot, Swaledale, Welsh Mountain) and a long-woolled ram (Border Leicester, Blue-faced Leicester, etc). In some areas, such as the Welsh/English Marches, pure-bred sheep, for example, Clun, Kerry Hill, are to be found. Upland sheep have a much greater potential than is realised in practice and, because of the small farm size in these areas, there is a great need to increase production and income.

Our research programme on upland sheep commenced from an entirely different baseline to that of hill sheep. Over the years a considerable amount of work on crossbred sheep in relation to nutrition, reproduction and grazing management had been done in various institutes and universities in this country. This allowed the structure of systems, clearly defined on the basis of operational decision rules, and thus the possibility of a comparison of systems on an experimental basis. The expectation was that, given extensive monitoring, those components which were less well understood would be revealed and subsequent component research would be more effectively determined. There have therefore been established parallel programmes of systems research and component research, the expectation being that more rapid, effective and comprehensive understanding is achieved.

Calves from suckler cows contribute a significant proportion of the income of upland farms. In practice wide variation exists in the management of beef suckler cows. In some areas producton of small weaned calves which have to be sold in the store market has become increasingly less profitable. Problems of spread of calving, proportion of barren cows, the production of calves of acceptable weaning weights, and the effective and efficient use of resources, need to be investigated. Although valuable work was being undertaken elsewhere, it was considered that there was a need for intensive investigation of the nutrition of the beef cow at different physiological phases as it affects milk yield, reproductive performance and use of body reserves, as well as an understanding of herbage utilisation.

Thus, two chapters (7 and 8) deal with relatively recent work where accumulated knowledge, although valuable, is inadequate and experimentation will require to continue for several years before the ideas about modification to existing systems or development of new ones can be applied.

Red deer

The Strathfinella Improvement Society, comprising scientists from both HFRO and the Rowett Research Institute, have collaborated in a project to study the feasibility of domesticating the red deer (*Cervus elaphus*). The initial phase of this project (1971-74) was the subject of a book "Farming the red deer" published by HMSO (Blaxter, Kay, Sharman, Cunningham and Hamilton, 1974). Another, reporting on recent work, is currently in preparation. Enough is now known about the management and potential of red deer for the commencement of a development programme concerned primarily with meat production under both hill and upland conditions.

Collaboration

On moving to the new headquarters on Bush Estate, HFRO became a member of the Edinburgh Centre of Rural Economy, an association of teaching and research establishments, and this improved opportunities for collaboration. In this context we have received support and assistance from the Edinburgh School of Agriculture, the Animal Breeding Research Organisation, the Animal Diseases Research Association, the Scottish Institute for Agricultural Engineering, the Scottish Plant Breeding Station and the Edinburgh Regional Computing Centre. The statistical expertise and advice required has always been provided by the ARC Unit of Statistics, now centred in Edinburgh, and this association we believe has been of mutual benefit. Close links have also been maintained with the Macaulay Institute for Soil Research, the Rowett Research Institute, the Grassland Research Institute, the Hannah Research Institute, the Welsh Plant Breeding Station, the Royal (Dick) School of Veterinary Studies and the Centre for Tropical Veterinary Medicine.

Throughout 25 years the support and encouragement of the Department of Agriculture and Fisheries for Scotland and the Agricultural Research Council have played a vital role in the development of the research competence and programmes of the Organisation.

The purpose of the Organisation has always been to investigate the problems of the agricultural use of hill land and, more recently, of upland areas. This implies, in the long run, the investigation of methods of securing increased conversion of natural and improved pastures into animal products, mainly meat and wool. It has always been accepted that these aims be consonant with the maintenance or improvement of the natural resource.

Since hill land in particular is used for a variety of other purposes such as forestry, recreation and conservation of wildlife, and is valued for its scenic beauty, HFRO has maintained a dialogue with, and indeed has collaborated with, institutions outwith the strictly agricultural field. Acknowledgement is made to the Highlands and Islands Development Board, the Forestry Commission, the Institute of Terrestrial Ecology, the Nature Conservancy Council, the Centre for Agricultural Strategy and the Red Deer Commission.

It being our primary concern to assist the agricultural industry, we have been especially grateful for the interest shown in our work and the active promotion of it by the agricultural advisory services — the Agricultural Development and Advisory Service in England and Wales, and the three Scottish Colleges of Agriculture, and also the Meat and Livestock Commission.

Staff and facilities

In the early years of the Organisation laboratory facilities appropriate to the field work were developed on the research stations. As research work became more detailed a need arose for suitable chemical support laboratories. Some of these were established in accommodation made available by the Animal Breeding Research Organisation. However, the limited facilities and the scattered laboratories were inadequate for the development of a scientifically based programme and planning of new centralised facilities commenced in the early sixties. In 1973 the Organisation moved into a comprehensive set of new buildings at the Edinburgh Centre of Rural Economy on Bush Estate. These buildings provide modern laboratories, animal accommodation, glasshouses and associated buildings suitable for a comprehensive range of basic and applied projects to be developed in the two departments of the Organisation — Animal Production and Nutrition, and Plants and Soils.

The biochemical and inorganic chemistry laboratories are provided with a range of modern equipment and an increasingly wider range of routine analyses has become available to scientists within the last 10 years. New technique development is also a feature so that more detailed and comprehensive interpretation of experiments has become possible. The range of experimental opportunities has been significantly increased, especially in the controlled laboratory, glasshouse and animal house environments as a basis for the more applied field research.

From a staff of 19 scientists in 1954 numbers increased to 47 in 1967 and to 84 in 1977, the greater numbers being due to an expansion in the support staff in the laboratories at Headquarters.

References

DOAS (1951). *Hill Farm Research*. Report of the Scottish Hill Farm Research Committee. Edinburgh, HMSO.

EEC (1975). Council directive on mountain and hill farming in certain less favoured areas (75/268/EEC)·*Off. J. EEC*, L128/1.

2

Soils and vegetation of the hills and their limitations

Contributors: M J S Floate, S A Grant, J King

Co-ordinator: P Newbould

Relationships between soil and vegetation

One of the first objectives of the Organisation's research programme was to understand the ecological status of the indigenous hill vegetation, its origin, and its relationship with soil type, and with grazing and burning managements.

Soil surveys were made by the Macaulay Institute for Soil Research at all the Organisation's research stations, and at two of these — Glensaugh and Sourhope — vegetation studies were also carried out. These were concerned with the relationships amongst vegetation, soil type and biotic history (Nicholson and Robertson, 1958: King, 1962). At Sourhope, where grassland rather than heather predominated, it was possible to distinguish between differences in pasture composition associated with both soil and boitic factors. In 1962 McVean and Ratcliffe published the results of a vegetation survey of the Highlands. This covered a wide variety of vegetation in the forest, sub-alpine and montane zones, the first two lying below about 600 m altitude and being of importance to pastoral farming.

To complement this and to provide a working classification applicable to Scotland as a whole, a survey was carried out in 1962-63 of forest and sub-alpine zone vegetation in south and mid-Scotland. For this a classification was adopted based on soil type and the results were published along with a condensed version of McVean and Ratcliffe's results in "The vegetation of Scotland" (King and Nicholson, 1964). A subsequent detailed study of the changes in pasture composition associated with changes in soil base status (pH) and drainage (Rogers and King, 1972a) provided further information on the vegetation/soil relationships used in the original classification.

The forest and sub-alpine zones contain a variety of vegetation types which can be related to three environmental gradients. These are variation in soil nutrient level (of which soil pH is a useful index), variation in soil drainage class and variation associated with biotic history, for example grazing pressure and frequency of burning.

In theory, for any given grazing pressure, there should be a range of vegetation types each growing on a soil of different pH and drainage class. Each vegetation type should be distinguishable by botanical characteristics which arise from the different soil conditions. However, vegetation on different soils has seldom been subject to the same biotic history so that, in practice, differences due to soil factors are confounded with differences due to biotic factors.

Each vegetaton type attracts a different grazing pressure and seasonal grazing pattern (Hunter, 1962) which exert a direct influence on botanical composition and have an indirect effect on soil development. Low grazing pressures lead to selective grazing; shrubby and other avoided species remain uneaten and tend to become dominant. High grazing pressures lead to unselective grazing and the predominance of prostrate species. Grazing pressures also determine how much of the vegetation is eaten and returned to the soil as excreta and how much decays *in situ* or is burned. Since these processes influence soil development it follows that soil, vegetation, grazing and decomposition are all interdependent. The vegetation is determined by soil conditions, modified by grazing and burning (Chapter 5) which, in turn, affect soil development. The result of this process carried on over several hundred years is to produce a mosaic of soil/vegetation ecosystems each contributing in a characteristic way to the food supply of the grazing animal.

The general relationships between hill soil and vegetation type are summarised in Table 1. Variation is related to soil base status and to soil drainage and the emphasis is on differences of agronomic importance. On the least acid soils above about pH 5.3 (mull soils) white clover (*Trifolium repens*) is present, the vegetation is a high grade *Agrostis-Festuca* grassland with wet-land species abundant on the poorly drained sites. Decomposition processes are rapid on these mesotrophic sites and, except where wet-land species are abundant, grazing pressures are high.

Table 1. Summary of the main soil and vegetation types of the hills

Soil	pH	Vegetation type	Principal species
Brown earth freely drained	5.3-6.0	*Agrostis-Festuca* grassland high grade or spp. rich	*Agrostis tenuis* *Festuca rubra* *F. ovina* *Poa* spp. *Trifolium repens* Herbs abundant
Gleys poorly drained	5.3-6.0	As above with wet-land spp. *Carex, Juncus*	*Carex* spp. *Juncus* spp.
Brown earth freely drained	4.5-5.2	*Festuca-Agrostis* grassland low grade or spp. poor	*Agrostis* spp. *Festuca ovina* *Pteridium aquilinum*
Gleys poorly drained	4.5-5.2	As above with *Nardus* and wet-land spp. *Carex, Juncus*	*Nardus stricta* *Carex* spp. *Juncus* spp. *Deschampsia caespitosa*
Podsols Peaty podsols freely drained	4.0-4.5	*Nardus or Deschampsia/* Festuca grass heath or *Calluna* shrub heath	*Nardus stricta* *Deschampsia flexuosa* *Festuca ovina* *Calluna vulgaris* *Vaccinium* spp. *Erica* spp.
Peaty gleys poorly drained	4.0-4.5	*Molinia* grass heath or *Calluna/Molinia* heath	*Molinia caerulea* *Festuca ovina* *Deschampsia flexuosa* *Calluna vulgaris*
Deep blanket peat poorly drained	3.5-4.0	*Trichophorum/Eriophorum/ Calluna* bog	*Trichophorum caespitosum* *Eriophorum* spp. *Calluna vulgaris* *Molinia caerulea* *Sphagnum* spp.

On soils between pH 4.5 to 5.0 low grade or species-poor *Festuca-Agrostis* grassland predominates. Clover is absent, decomposition rates are slower and humus tends to accumulate at the soil surface. Grazing pressure is moderately high except where avoided species, such as bracken (*Pteridium aquilinum*) are abundant.

On still more acid, podsolic or gleyed soils with pH less than 4.5 (mor soils) either shrub heath or grass heath are found depending on the past grazing and burning history. Avoided species such as *Molinia caerulea* or

Nardus stricta predominate in the grass heaths and humus tends to accumulate as peat. The more poorly drained heaths have affinities with bog vegetation which is characteristic of the deeper and more acid peat soils.

Distribution and properties of hill soils

Distribution

Although comprehensive studies of the evolution and distribution of soil types in hill areas have never been part of the Organisation's research programme and are more properly the function of soil survey organisations, Floate (1977a) has described the topographic distribution of hill soils and discussed some of their properties. The distribution is primarily related to altitude and, for a given parent material, ranges from poorly drained gleys in the valley bottoms through brown earths on the lower slopes to podsols and peat formation above 600 m. A description of the distribution and pastoral capability of major soil groups in west Scotland has been prepared following collaboration between the Macaulay Institute for Soil Research and HFRO.

Properties

The relationships between the properties and the problems of hill soils have been discussed by Floate (1977a) and Newbould and Floate (1979).

An important feature of the topographic soil sequences mentioned above, is the increasing accumulation of organic matter — either within the soil profile or on the surface as peat or humic material — which occurs with greater elevation. The increasing inorganic matter content and associated cation exchange capacity, together with stronger leaching and the loss of bases results in progressive base unsaturation and the development of acidity associated mainly with exchangeable aluminium in brown podsolic soils and podsols, and with exchangeable hydrogen in the more organic soils (Floate, 1978b).

The accumulation of organic matter has important effects upon the availability and distribution of the nutrients required for plant growth. The reputed 'low fertility' of hill soils is caused by the slow rate of decomposition of accumulated organic matter leading to a slow release of nutrients by mineralisation, nitrogen being particularly affected. A further consequence is the fixation of phosphorus associated with acidic weathering

Agrostis-Festuca grassland

Nardus grassland

Calluna shrub heath

Trichophorum/Eriophorum/ Calluna bog

Pasture sampling on indigenous hill vegetation

products, like the oxides of iron and aluminium in the podsols and brown podsol soils. This does not mean that the total amount of plant nutrients is necessarily small or lower than in lowland soils (Floate, 1977a) but rather that the amounts readily available for plant growth are small, and that the rate of replenishment is very slow. When little or no fertilisers are applied to grazed hill pastures these factors become critical for the quantity and quality of herbage produced.

The significance of phosphorus in the accumulation of soil organic matter in permanent grassland soils has been demonstrated in both Britain and New Zealand. The limitations imposed by the availability of inorganic phosphorus are important in this process and influence the nature and composition of organic matter in more strongly leached soils. It has been suggested that what is required to improve such soils is a return to conditions more typical of an earlier stage in soil formation, and Floate (1970d, 1973) has shown the importance of intensified grazing in this process. Some aspects of the philosophy, principles and practice of hill land improvement will be considered in detail in Chapter 3.

The cycling of nutrients from the soil, through the plant and the grazing animal, back to the soil is important in the maintenance of soil fertility as indicated by King and Nicholson (1964). This process has since been studied in detail by Floate and is discussed in Chapter 5.

Limitations to pasture production

Output from hill and upland farms is dependent upon the interaction of a number of factors which have been appraised by Eadie (1970b). These include pasture production, pasture utilisation, animal nutrition and management considerations. The limitations to herbage production, whether of native or improved pasture, involve both environmental and soil factors.

Environmental

At the highest elevations climatic factors impose overriding limitations on pasture production but at intermediate and lower levels the relationships between altitude and climate, climate and soil, and soil and vegetation variously limit the productivity of pasture types so that potential production may vary greatly from site to site. In general, as altitude increases the climate becomes more severe, soil nutrient availability decreases and plant growth is reduced. The cool, oceanic climate of the United Kingdom has winter temperatures limiting to growth and the seasonal amplitude in mean monthly temperature is small. This has the result that small increases in altitude are accompanied by relatively large effects on micro-climate and particularly on the length of the growing season.

The effect of altitude. In 1961 when research was being inititated on role of land improvement and on problems related to the reseeding of hill pastures it was considered that a knowedge of the limitations to growth attributable to the climatic component of altitude would be of value (HFRO 3rd Report). Such knowledge would assist in setting the context, aims and objectives of future research aimed at overcoming limitations set by soil conditions.

Sites to study the climatic component of altitude were chosen and were set up to eliminate variation in slope, aspect and soil type. Clonal material of *Lolium perenne* (S.23) was grown at 33 sites and with altitude ranging from 150-680 m; biological and meteorological data were collected over a period of four years from 1962-65.

Yield over the year as a whole diminished by 2% and floral development of the grasses was delayed by 1.3 days for each 30 m rise in elevation. The magnitude of the effect on yield, however, varied with season. In spring, when temperature is limiting, yields were decreased by some 5% for each 30 m rise in altitude but in autumn by only 1.8%. In summer, yield differences were non-significant or trends reversed, the highest yields often occurring at intermediate or highest altitudes. This result was related to the development of moisture stress at the lower altitudes (Hunter and Grant, 1971).

A similar study by King, Grant and Rogers (1967) using indigenous *Festuca rubra* growing alone and in mixture with *Trifolium repens* also showed yield reductions in spring of the order of 5% for each 30 m rise in altitude. Over the season as a whole highest yields were found in the altitude range 300-380 m. Moisture deficits occurred regularly in summer at the lowest altitude (230 m) but were rare above 460 m.

On some of the altitudinal transects, exceptionally exposed sites at low or intermediate elevations allowed assessment of the importance of wind exposure effects. Effects on yields during wet periods of the year, which were also usually the cooler periods, were of minor importance compared with the effect of lowered temperatures associated with altitudinal increase. In dry spells, however, when soil moisture stress was limiting, yields from exposed sites were significantly depressed and were inversely related to degree of site exposure rather than altitude (Grant and King, 1969).

Temperature and spring growth. Whilst these and similar studies by other workers indicated that high yields are possible in the hills following land improvement, they also indicated a significant delay to the start of growth in spring with increasing altitude. Lambing of hill sheep takes place in April and in many years when spring is late there is a shortage of grass on reseeded areas. This problem is aggravated on hill reseeds and it was felt that there was a need for grass varieties with improved production at temperatures below 10°C.

Some work was carried out during the period 1965-68 to examine the inter- and intra-specific variation in various growth parameters of a range of hill grass species under winter conditions (Grant, 1968b). Facilities were very limited. Temperatures were monitored at sward height and measurements were made during successive growth periods of rates of leaf extension, leaf appearance and tillering on material grown in boxes in specially constructed outdoor cabinets. Variation in temperature accounted almost entirely for variation in leaf extension rates within clones. Rates of leaf appearance were closely dependent on temperature during cold weather. In milder spells, however, shading depressed leaf appearance. Rates of tillering were affected by both temperature and light although, in general, shading had a larger effect at higher temperatures. Inter- and intra-specific variation were sufficiently large to suggest that, with plant selection and breeding, it should be possible to isolate varieties of grasses with improved yields at low temperatures. *Festuca rubra* provided some particularly promising material. Work in this area was discontinued in 1968 because the facilities available for such studies did not permit their extension and because similar work was being carried out at the Welsh Plant Breeding Station. The decision was also influenced by the need to expand activities in other areas considered to have higher immediate priority (Chapters 3 and 4).

Excessive soil wetness. Another aspect of climatic limitations in the hills and uplands is excess wetness which may result from the interaction of rainfall, topography and soil drainage characteristics. Studies were made of the abundance and distribution of grassland species in hill pasture in relation to soil wetness or aeration and soil base status (Rogers and King, 1972a). Consistent species distribution patterns were found in relation to the variation of the two factors studied. It could not be assumed, however, that the relationships between the edaphic factors and species distribution patterns were causal as the patterns were as much a product of competition from other species as an expression of specific ranges of environmental tolerances.

Performance of herbage species in relation to soil moisture status and aeration in the absence of competition was examined in further work using sown grass species (Rogers and Davies, 1973). Of the species studied *Dactylis glomerata* showed the greatest reduction with respect to yield under waterlogged conditions. In all species dry matter yield, potassium, nitrogen and magnesium content were positively correlated with soil oxygen concentration. Calcium concentration in the plant tissues showed no relationships with soil moisture or aeration but in *D. glomerata* it was related positively with soil calcium concentration.

Work was also carried out in the early years in conjunction with the Macaulay Institute for Soil Research to investigate the hydrology of peat catchments (Robertson, Nicholson and Hughes, 1963). The relationships amongst rainfall, run off, evapotranspiration and fluctuations in the water table were studied over a three-year period at a raised bog in Lanarkshire. Monthly run off as a percentage of rainfall was consistently high in the months of November, December, January and February, and consistently low in June. In other months there was great variation from year to year. Similarly, the height of the water table was most variable from March to October.

This variability no doubt contributed to the large year-to-year variation in the vigour of growth observed in recent studies among species during annual monitoring of the floristic composition of grazed plots on blanket bog (Grant, unpublished). The dry summers of 1975 and 1976, for example, were followed in 1977 by unusually vigorous growth of heather (*Calluna vulgaris*) and reduced growth of *Eriophorum angustifolium*.

Soil

Physical factors. The influence of the nature of the parent rock on the distribution of soil types and their properties was briefly considered by Floate (1967, 1977a). This affects soil depth, stoniness and texture as well as the topography, all of which can be serious limitations on both existing levels of production and potential for improvement. Whilst it is recognised that these limitations exist, it is also clear that they are of a kind which cannot readily be changed by improvement techniques. This is the main reason why no work was carried out in HFRO on the effects of these physical factors on pasture production.

Chemical factors — acidity. Greater attention has been given to the nature of chemical limitations in soil fertility, and to methods for their estimation and correction. Floate (1967) suggested that soil factors are probably more important in limiting production on potentially improvable soils, than are climatic factors, and Floate (1977a) considered these in relation to other factors, and identified acidity and nutrient supply as the major chemical limitations.

Acidity is almost certainly the most common, and frequently the over-riding, chemical aspect of soil limitation on pasture production. Early work

Determining
botanical composition
of pasture

Measuring
photosynthesis,
respiration and
nitrogen fixation

Measurement of nitrogen fixation
using acetylene reduction method

Experiment in
controlled environment

on decomposition of organic matter in hill soils showed acidity as a major factor controlling decomposition rate, and as shown by Floate (1970d) this rate may be the limiting factor to the cycling of nutrients in the soil-plant-animal system. The importance of this when available nutrients are limited, and when the cycle operates in a nearly closed system will be discussed in Chapter 5. The extent of acidity limitations in the north of England was illustrated by work from the University of Newcastle which showed that lime and fertilisers raised herbage dry matter yields from about 2000 kg/ha to 10,000 kg/ha from grass/clover pastures sown on reclaimed land. With lime alone, yields of up to 6000 kg DM/ha were achieved. It is likely that this response to the removal of acidity limitations was due primarily to stimulated organic matter decomposition and the consequent release of available nutrients, shown to be an important process in hill soils (Floate 1970a, d, 1977a).

Difficulties arise when assessing the lime requirement of hill soils and these are discussed in Chapter 3.

Chemical factors — nutrient supply. The high levels of organic matter in hill soils are important with respect to their reputed 'low fertility'. The latter is usually due more to the small amounts of nutrients available for plant growth and the slow replenishment of this supply, than to an absolute shortage of nutrients. Nitrogen and phosphorus are thus largely present in the rooting zone of hill soils in organic forms which must be broken down to inorganic forms by mineralisation before they become available for plant growth. Severe limitations on plant growth due to deficiencies in available nitrogen and phosphorus have been appreciated by many workers although different approaches have been taken to their correction. Responses in pasture production and improvement in quality can be obtained by the enhanced mineralisation of nutrients from organic matter, especially from nitrogen and phosphorus, and by clover-fixed nitrogen stimulated by the use of small amounts of fertiliser. These aspects are discussed more fully in Chapter 3.

It seems that established pastures (either *Agrostis-Festuca* or reseeded ryegrass/clover) respond to additional supplies of nitrogen which may come from fertilisers, clover fixation, or animal excreta (Floate, 1970d), whereas phosphorus supply may be more important for establishment. In order to use phosphorus fertilisers efficiently to supplement the soil phosphorus supply, which may be enhanced by mineralisation from soil organic matter and animal faeces, it is necessary to determine the current phosphorus status of a soil and to know the response of reseeded pastures to added phosphorus on the same soil. Tests of methods to achieve this are also discussed in Chapter 3.

Recent studies have shown that soil aluminium and phosphate sorption are higher in soils on basalt than on other brown podsolic soils (Mitchell and Floate, unpublished). These basalt soils fail to respond to added phosphorus in the absence of lime. Aluminium appears to limit plant uptake of phosphorus in acid soil conditions but the mechanism of this process is not yet fully understood.

Interactions between acidity and potassium supply have also been observed on peat. In pot experiments with clover, added potassium gave large responses in yield which were enhanced by added phosphorus (Rangeley and Newbould, unpublished). Potassium deficiencies at low lime levels were also diagnosed on peat at Lephinmore (Floate, Rangeley and Bolton, unpublished) and the importance of both applied and recycled potassium is discussed in Chapter 3.

These and other studies have demonstrated that responses can be obtained from additions of major nutrients and that their shortage is only a temporary limitation to existing levels of herbage production. Quantity of herbage is not the only limitation to sheep production because pasture quality and mineral nutrient content are frequently low in indigenous pastures. Both pot and field experiments have indicated that the mineral content of herbage from untreated soils is low, and that the plant content of these major nutrients can be significantly increased when their availability in soil is enhanced either by fertiliser additions, nitrogen fixation, mineralisation of organic matter, or recycling of nutrients in animal excreta (Chapter 5).

Although minor elements probably do not often impose significant limitations on plant growth at current levels of intensification and herbage production, some indications of deficiencies were obtained on peat at Lephinmore. Significantly depressed yields of clover in the absence of added molybdenum, and of ryegrass in the absence of added boron were recorded in some years (Floate, unpublished). Low levels of copper and cobalt and/or high levels of molybdenum in plant material are, however, of greatest importance in animal nutrition.

Indigenous pasture — limitations to its utilisation

The seasonal patterns of growth and senescence affect the quantity and quality of herbage available for grazing. The native vegetation types of the hills and uplands are characterised by low dry matter production of moderate to poor quality herbage.

Hill species fall into the following three categories of seasonal growth rhythms:—

(a) Species capable of growth at any time of year provided temperature and light levels are suitable. Most acid grassland species fall into this category.

(b) Species capable of growth during a restricted growing season, at least in terms of shoot extension, but remaining winter green. Examples of species in this group are provided by the heaths.

(c) Species capable of growth during a restricted growing season and which die back in autumn thus providing no winter grazing. Most species in this group are found in wet habitats and examples are *Molinia caerulea, Trichophorum caespitosum* and some *Juncus* species.

The net annual dry matter production and digestibility of the pasture types described at the beginning of this chapter are summarised in Table 2. The quality of the herbage is greatly dependent on management. This is because regrowth is of higher quality than ageing, ungrazed material. An indication of the magnitude of management effects on quality is included in the table by quoting indices for both unutilised pasture and regrowth in autumn.

Limitations to the utilisation of pasture inherent in the plant are two-fold in origin. Where the quality of the herbage is good the amount of pasture and its growth capabilities are limiting. On the other hand, where quality is poor, the nutritional penalties to the grazing animal limit the pattern and extent of their use. More will be said about the consequences of given levels of nutrition to animal performance in Chapter 4 and the inter-actions between pastures and grazing animals in Chapter 5. Of the native vegetation types, only the acid grassland and related types fall into the category where, with good management, quantity is often more limiting to animal production than quality. In all other communities, despite low levels of herbage production, the quality of herbage is the major limiting factor.

Some management possibilities

In summary, hill land production limitations are due to five groups of factors — climate, site, soil, vegetation and management. Floate (1977a) pointed out that what might be regarded as permanent limitations due to temperature, rainfall, altitude, slope etc. are predominant in the first two groups. Some local modification of climate might be provided by shelter, and excess wetness due to topography can sometimes be improved by drainage but the cost is often prohibitive. Some soil limitations, for example those due to stoniness, texture and shallow depth are permanent but there is a larger number of factors, such as those due to acidity and nutrient shortage, which can be corrected. With factors related to vegetation and management the more permanent limitations are of minor import-ance and improvements can be made. Whilst the removal of most limitations due to existing vegetation, such as quantity and quality of herbage, is technically possible, the cost/benefit relationships of their removal

Table 2.   Dry matter production and quality of native hill pastures

| Vegetation type | Dry matter Yields, kg/ha | Basis | Quality index — digestibility as a percentage of DM | | | |
			May-June	September 1st cut	September regrowth	January-March
Agrostis-Festuca (acid grassland)	2200[1] 3200	sum, monthly cuts total green DM	70-76[10]	45-55	65-73	40-50
Festuca-Deschampsia (grass heath)	2240[2, 3]	sum, monthly cuts	65-75[10]	50-55	(65-73)	—
Nardus (grass heath)	1000-2240[1, 2] 4000[10]	sum, monthly cuts total green DM	60-70[10]	45-50	—	35-40
Molinia (grass heath)	1110[3] 1700[4] 2000-4000[5]	sum, monthly cuts green DM April-Aug total standing crop	60 65-70[10, 11] —	— 45-50 —	— — —	— (38-litter) —
Calluna (shrub heath)	1750-3400[6, 7, 8]	current shoots heather cover 85%+	60[10, 12]	50	55	40
Trichophorum/Eriophorum/ Calluna (blanket bog)	1450-1700[8, 9]	current shoots heaths + green material grasses and sedges	60-68[11]	40-55	—	—

Sources:
1. Floate, 1970, *J Br Grassld Soc* 25, 295-302
2. Nicholson, HFRO, unpublished
3. Milton, 1940, *J Ecol*, 28, 326-356
4. Grant *et al*, 1963, *J Br Grassld Soc* 18, 249-257
5. Loach, 1968, *J Ecol* 56, 433-444
6. Grant, 1971, *J Br Grassld Soc* 26, 173-181
7. Miller and Miles, 1969, *Grouse research in Scotland, 13* report NCC Banchory pp 31-43
8. Grant, HFRO, unpublished
9. Forest, 1971, *J Ecol* 59, 433-479
10. HFRO, unpublished
11. Grant and Campbell, 1978, *J Br Grassld Soc* 33, 167-173
12. Milne, 1974, *J Agric Sci, Camb*, 83, 281-288

require evaluation. The study of the possible benefits of bracken control is an example of such work (Chapter 3). The final group of limitations due to management factors is discussed in Chapters 4, 5 and 6.

While some of the earlier work at HFRO attempted to evaluate the extent of permanent limitations, most recent work has been concentrated on the temporary limitations due to soil, vegetation and utilisation factors. It is the means of reducing these limitations and the techniques of land improvement which are discussed in detail in Chapter 3.

Reference

McVEAN, D N and RATCLIFFE, D A (1962). *Plant communities of the Scottish Highlands: a study of Scottish mountain, moorland and forest vegetation.* Nature Conservancy Monograph No. 1. London, HMSO.

3

Establishment and maintenance of sown pastures

Contributors: G E Davies, M J S Floate, A Haystead, J King, J A Rogers
Co-ordinator: P Newbould

Introduction

As indicated in Chapter 1, at the time HFRO was formed, techniques devised in Wales by Stapledon, Davies and Ellison for changing or improving indigenous hill pastures by the introduction of more productive and nutritious plants were already well known. These depended largely on the use of deep ploughing to destroy and bury the existing vegetation and although successful, except in wetter areas where buried rush seeds were encouraged to germinate, the approach was largely seen as a means of extending the area of in-bye land. The initial botanical work at HFRO was directed primarily at defining the resources as described in Chapter 2, and at understanding the impact of the grazing animal on the existing hill vegetation and its successions with a view to determining how best to manage hill pastures within the context of the traditional system (Wannop, 1958b, Hunter, 1958b, Nicholson, 1964a). A detailed description of the further development of this theme forms the basis of Chapter 5 and is mentioned only briefly in this chapter. The main emphasis here is on other methods of pasture improvement, particularly the establishment and maintenance of sown pastures.

Despite the early concentration on hill pasture improvement by grazing control, which is the cheapest method (Table 3), the need for more direct and rapid methods of land improvement was already clear and two main alternative approaches were investigated. One approach was to encourage beneficial changes in existing pastures by use of selective herbicides (King and Davies, 1963) or by surface applications of lime and phosphate (Robertson and Nicholson, 1961) which increase the rate of change but incur additional costs. The second strategy was to destroy the existing vegetation and to introduce other species of plants using surface seeding and the minimum of cultivation which is the most rapid but most costly method. Such procedures have evolved over a number of years and work at many places in addition to HFRO has helped to make them more reliable.

Table 3.   The main processes and relative cost of hill land improvement procedures used at HFRO

Fence to control grazing	Cost relative to fencing
Alone	1.0*
+ herbicide : dalapon to reduce *Nardus*	1.3
: asulam to reduce *Pteridium*	1.7
+ lime and phosphorus	2.1
+ lime, phosphorus and white clover seed	2.2
+ lime, phosphorus, potassium, starter nitrogen, white clover and grass seed	
by oversowing	3.2
with light cultivation (rotovation/rolling)	3.7
with ploughing	4.4
+ tile drainage system	7.6

* The base line was taken as £85, i.e. the gross cost per ha of enclosing an area of 8 ha with a mains electric fence costing 60p/m.

The possibility of using aircraft to spread lime and phosphate, as in many parts of New Zealand, was explored but found not to be a viable proposition at that time (Nicholson, 1959a, b, 1960a). A scheme to test the idea that soil fertility, and hence pasture improvement, would spread out from small areas of improved pasture placed in a larger area of unimproved hill land was set up under field conditions at Lephinmore (Nicholson, Currie, Patterson and McCreath, 1968). With the vegetation type at this Research Station marked responses were not evident. The transfer and recycling of nutrients through the grazing animal, however, has been shown to be vital to the continued growth of improved hill pastures (Chapter 5). The choice of grass species and varieties to establish and grow well in the difficult conditions on the hills was explored by Hughes and Nicholson (1961/2). The results of the latter investigation led to studies of the climatic limitations to the survival and production of plants on the hills, and the effects of utilisation in such conditions (Chapter 2)

This work resulted in a better appreciation of the short growing season of hill vegetation. It was noted, however, that in the short period of growth the quality of the herbage could be as high as that of lowland pastures (Eadie and Black, 1968). It was this information, plus the accumulated knowledge of the nutritional requirements of hill sheep (Chapter 4), that led eventually to the formulation of what has become known as the 'two-pasture system' (Eadie, 1970a).

Following this formulation the emphasis in land improvement studies switched to the investigation of the more rapid responses that follow sward destruction and costly inputs of lime, phosphate, seeds and cultivation (Table 3). It should be emphasised that the more traditional and slower changes in hill vegetation brought about by grazing control alone still have a role in certain circumstances, and irrespective of the precise methods of pasture improvement adopted, the initial and fundamental requirement in all circumstances is the provision of fencing to allow control of grazing.

Although the general requirements to relieve the temporary limitations of acidity, low soil fertility and poor quality plant material in hill situations have been known for some time (Chapter 2, Newbould, 1974/5, 1979) there was a lack of quantitative information over the whole range of soil types on the economically justifiable changes needed to initiate and maintain an improvement scheme. The pressure to quantify the biological and production responses of soil amelioration and alteration of plant material in the hills was further increased by the rising costs of transport and raw materials, the fluctuating subsidy and produce price position, and the cost of borrowing finance. Thus, the assessment of the cost/benefit of land improvement procedures has become uppermost in farmers' and advisers' minds and the relevant calculations can only be made if the biological responses are known.

To meet this situation a series of laboratory, glasshouse and small plot field experiments commenced in the seventies as new facilities were provided and much of this work is still in progress.

Difficulties arise in translating results obtained by growing plants under glasshouse conditions in finely milled and uniformly mixed soils, to field conditions. On the other hand, experiments in the field, particularly with grazing animals, have to be small to keep costs within reasonable bounds. This raises difficulties of obtaining reproducible results because replicate areas of uniform hill soil or vegetation are extremely hard to find. Moreover, field experiments have to be repeated in several years to account for variation in climate.

The need to choose suitable species of pasture plants, especially legumes, to use herbicides, to prepare seed beds, to add lime and fertilisers both to establish and maintain improved pastures is common to all the procedures where indigenous plants are replaced and the HFRO contribution is described in this chapter. It concludes with a brief examination of the responses in pasture production and quality to be expected from the introduction of sown species and indicates some areas where information is still lacking. Before commencing a detailed description of work on these procedures attention is drawn to the following cautionary points. Because of the high cost of any scheme of pasture improvement and because both the initial and maintenance costs vary enormously between different sites, the first essential when planning a land improvement scheme is to be certain that the most responsive, accessible site has been chosen and that the additional herbage can be utilised fully and profitably within the existing or a modified farming system (Chapter 6). Moreover, there must be sufficient animals to use the extra herbage and adequate finance to provide maintenance dressings of lime and fertiliser to prevent the improved pasture deteriorating rapidly. There must also be awareness of the essential role of fencing and the possible occurrance of veterinary problems due to movement and concentration of stock and to trace element problems arising from the effects of lime and change of plant species.

Choice of pasture plants for the hills

The need for more rapid changes in herbage quality and production than could be achieved by grazing management alone stimulated interest in the introduction of sown species. The main emphasis was to establish a legume, usually white clover *(Trifolium repens)*, together with companion grasses that could benefit from the nitrogen fixed and transferred by the legume. The requirements for these two classes of plant are considered separately here. Although work has continued in parallel the main emphasis at HFRO has been on white clover. Other organisations in the Agricultural Research Service have greater responsibility for work on the grasses.

Legume

Deficiency of available nitrogen is a key limitation to the growth of hill pastures. Unlike lime and phosphate which are effective for several years after application, dressings of nitrogen are soon lost from the soil and there is a need for continuing applications. The annual application of fertiliser is inconvenient for hill farmers and is becoming ever more expensive. Even some upland and lowland farming systems may soon be unable to bear the cost of its regular use. Thus, the emphasis has been and still is on plants, such as legumes, that can provide their own nitrogen. Within the legume family, white clover is the only useful species which is able to tolerate the climatic and soil conditions of the hills to even a limited extent. Work on alternative legumes, such as species of *Lotus,* is continuing elsewhere but has not yet produced material with the dry matter production and quality to match white clover. The potential value of white clover has always been appreciated at HFRO but the emphasis in the early seventies on improvement by complete change of vegetation led to an upsurge in interest.

Microbial requirements of white clover. The role of white clover in hill land improvement and the biological reasons for its importance were discussed by Newbould and Haystead (1978). The process by which bacteria of the genus *Rhizobium* infect the roots of white clover and form nodules in which nitrogen from the atmosphere is fixed symbiotically, was described in some detail. The early work of Holding and King (1963) and Singer, Holding and King (1964) indicated that not all hill soils contained the appropriate strains of *Rhizobium* in sufficient numbers to form an effective symbiosis. Laboratory studies here and elsewhere have shown that the introduction with the seed (inoculation) of organisms which match the clover to be grown bring about considerable increases in seedling establishment and survival, in nitrogen fixation and in growth.

When similar responses were obtained in a field experiment at Lephinmore on deep peat soil in 1972, it was decided to mount a more thorough investigation on a wider range of soil types. A collaborative project with the three Scottish Colleges, the ARC Unit of Statistics, the Microbiology Department of Edinburgh University, the Welsh Plant Breeding Station and three MAFF Experimental Husbandry Farms (Great House, Pwllpeiran and Redesdale) was set up to carry this through. Twenty-one, experiments were carried out on 15 soil types over a four year period. The results indicate that consistent responses to inoculation were observed only on deep peat soils and even on these soils the responses occurred only in the first year; contamination of the control or untreated plots occurred thereafter. There were intermittent or inconsistent responses in seedling establishment and plant cover on other soil types and these require further investigation. On many of the mineral soils where no agronomic responses were observed it was shown that the introduced strains of *Rhizobium* had formed nodules on the roots of the white clover plants. However, it appears that either there were too few good nodules to have an impact on plant growth or other factors were affecting the expression of their nitrogen fixing abilities.

Research to resolve these problems and inconsistencies is still continuing but it is the Organisation's view that all clover seed to be sown in peat soils must be inoculated with an appropriate strain of *Rhizobium*. Further, because the cost of this treatment is small compared with that of the seed itself and with the costs of other aspects of hill land improvement, it would be a reasonable policy for clover seeds sown in all hill soils to be inoculated.

More recently it has been established that in some soils there are benefits from increasing the level of infection of white clover roots with mycorrhizal fungi. It is likely that in this organism, as with *Rhizobium*, some strains may be more beneficial than others and the development of appropriate methods of inoculation are proving to be even more difficult than with *Rhizobium*. The mycorrhizal association apparently functions by effectively extending the root surface which enables more of the small amounts of available soil phosphate to be absorbed and used for the growth of plants and, particularly, of nodules. There is thus evidence of an additive effect when white clover forms symbioses with both *Rhizobium* and mycorrhizal fungi. A collaborative group to co-ordinate studies of the field side of the mycorrhizal aspect of the work has been set up involving Rothamsted Experimental Station, Leeds University, Dundee University, the Agricultural Development and Advisory Service and HFRO. Work on other aspects is in progress at most of the laboratories described above and in many other parts of the world but especially in New Zealand and Australia.

Nutrient requirements of white clover. It is well known that white clover grows poorly in acid soils and this is illustrated by the absence of this plant from all but the relatively high pH, well drained brown earth soils (Table 1). King (1961a) examined characteristics of indigenous populations of white clover which he found to be smaller leaved, more prostrate and of earlier flowering date than the commonly sown S.184. However, he considered these characteristics reflected a response to high grazing pressure on vegetation of more basic soils than to a direct relationship with soil base status. Thus, there is no inherent reason why white clover cannot grow well in hill soils provided it is placed in reasonably basic soil and supplied with adequate nutrients.

White clover requires at least a pH above 5.2 and has an optimum of about pH 6.0; however, it performs quite well at pH 5.5-6.0. To avoid problems due to lowered availability of cobalt or enhanced availability of molybdenum, however, pH 5.8 is probably the highest desirable level to have in hill soils following liming. White clover requires readily available phosphate and, in the presence of all other nutrients, responses have been shown to occur up to 320 kg P/ha (Rangeley and Newbould, unpublished). It is usual to apply between 60-80 kg P/ha. The plant also has a high demand for potassium. Laboratory and field experiments (Floate, Rangeley and Bolton, unpublished) indicated that about 120 kg K/ha were required on deep peat soils to balance the usual applications of phosphate. Responses to trace elements which are needed for satisfactory nodulation and fixation of nitrogen have been recorded but these are variable and not related directly to soils of a particular series or major soil group.

Some uncertainty surrounds the amount and time of application of fertiliser nitrogen to the seed bed for white clover and the amount of fixed atmospheric nitrogen that is transferred to companion plants (grasses) in the pasture. Work on both these subjects is currently in progress and a brief summary of the present position follows.

Interaction of mineral nitrogen uptake and symbiotic activity in the establishment and growth of white clover. Two sources of nitrogen are avaiable to the nodulated clover plant, inorganic (including fertiliser) nitrogen in the soil and atmospheric nitrogen (N_2). The rate of uptake of nitrogen from the soil is limited primarily by its availability (which in turn is a function of the amount present and the level of below-ground competition) and by factors affecting the growing shoot. The assimilation of atmospheric nitrogen on the other hand is additionally affected by the development history of the symbiosis and the availability of soil nitrogen. The effect of soil nitrogen on clover growth and symbiotic activity is modified by the age of the plant/*Rhizobium* association. Seedling growth in pots, for example, is accelerated by the application of inorganic nitrogen after nodule initiation has taken place, whilst the same level of nitrogen applied just before germination does not stimulate growth to the same extent and can even reduce growth by delaying the infection process (Haystead and Marriott, 1979a).

Field studies have been carried out at HFRO to determine the effect of rates and times of application of fertiliser nitrogen on symbiotic development of white clover and on clover and grass growth (Haystead and

Marriott, 1979a). As yet insufficient data have been produced to allow any general conclusions to be drawn but the results of a trial on deep peat have shown:

(a) No measurable response to a starter dressing of 30 kg N/ha applied at sowing time.

(b) Higher rates (90-120 kg N/ha) applied after nodule initiation depressed clover nodulation and nitrogen fixation but did not affect clover growth.

(c) On plots given 90-120 kg N/ha after nodule initiation grass growth still showed improvement in the second year whilst the depressant effect on clover nodulation and nitrogen fixation had disappeared.

A more fundamental question arises when the growth of the nodulated legume on N_2 or mineral nitrogen (NO_3^- or NH_4^+) is considered in terms of the energy relationships within a single clover plant. Contradictory views have been expressed — several workers regard the energy requirements of NO_3 assimilation and N_2 fixation to be identical at the whole plant level while others have shown respiratory carbon loss in subterranean clover (*Trifolium subterraneum*) and soybean (*Glycine max*) to be greater in plants utilising N_2. Using a technique developed at HFRO, it has been shown (Haystead, King and Lamb, 1979; Haystead, King, Lamb and Marriott, 1979) that the potential growth rate of nodulated white clover growing on 150 ppm NO_3, is about 10% greater than growth on N_2. In the short term, however, the difference in shoot growth (herbage production) was less than 10% because a smaller proportion of the available carbon was utilised in root growth in N_2-fixing plants.

Transfer of fixed nitrogen to non-fixing species. Techniques have become available over the last few years which enable the researcher to monitor nitrogen transfers within the soil/plant system in considerably greater detail. The acetylene reduction technique is a sensitive assay procedure for N_2-fixing activity which is invaluable in qualitative and comparative studies (see, for example Haystead and Lowe, 1977). Unfortunately, the only available radioisotope of nitrogen (^{13}N) has a half life of only 10.6 sec and is unsuitable in experiments in which transfers are quite slow. Although the naturally occurring stable isotope of nitrogen (^{15}N) is detected with 10^2-10^3 times less sensitivity than ^{13}N it has been used with considerable success at HFRO and elsewhere to demonstrate some of the intricacies of the below-ground ecosystem.

Using the acetylene reduction technique (Newbould and Haystead, 1978) and ^{15}N (Haystead and Marriott, unpublished) seasonal profiles of N_2-fixing and nitrogen-transferring activity in clover/ryegrass swards have been produced. Comparison of these profiles with meteorological data gives an indication of the environmental factors limiting N_2-fixation and the sensitivity of growth and N_2-fixation to defoliation throughout the year.

Sensitivity to defoliation in terms of loss of N_2-fixing activity (Haystead and Marriott, 1978) has been shown to be related to morphological adaptation of the clover plant in response to grazing pressure (King, 1961a; Haystead and Marriott, 1978) and to the carbohydrate status of the plant prior to defoliation (Haystead and King, unpublished). In general terms, however, in early spring, clover N_2-fixation is limited by low soil temperature and under some circumstances is slower to start than clover shoot growth. Once growth has become vigorous, and provided grazing pressure is not too great, N_2-fixation usually increases to a rate of 0.5-1 kg/ha/d until the separate or combined effects of flowering and moisture stress reduce fixation in early or mid-summer. During and after the mid-summer decline, extensive loss of root and nodule material occurred in grazed swards, rested and later cut to 1 cm on reaching a height of 4.5 cm (Haystead and Lowe, unpublished). During the spring period of vigorous growth, however, grazed swards showed little response to defoliation; cutting from 5.5 cm to 1 cm, for example, had no measurable effect on either N_2-fixing activity or nodulation (Haystead and Marriott, 1978).

A second peak of N_2-fixing activity occurs in autumn and is usually curtailed by decreasing soil and air temperature and finally by frost damage to the shoot. Although growth and N_2-fixing activity in autumn can be as great or greater than the spring maximum (Newbould and Haystead, 1978), nodule loss occurs throughout this period (Haystead and Lowe, unpublished).

At a more detailed level studies have commenced on the processes involved in the underground transfer of fixed nitrogen to companion grasses. In particular, information is being sought on the residence time of nitrogen in various pools within the soil (Haystead and Marriott, 1979b). Work is in hand to describe the role of the soil microflora in these processes and to determine how environmental factors affect the relative rates of mineralisation and humification of clover nitrogen and its immobilisation by microbes.

Variety of white clover. The aim of most hill land improvement schemes is to create conditions and to select sowing time and variety such that white clover will germinate, establish, form nodules and start fixing nitrogen as quickly as possible. In HFRO pasture improvement schemes, New Zealand Grasslands Huia (NZGH) at 2.25 kg/ha is currently the variety sown as it has a reputation for rapid establishment. However, it is also important that a significant white clover population should persist for as long as possible and since NZGH has a susceptibility to disappear quickly under United Kingdom conditions a proportion of S.184 or Kent Wild White Clover (0.75 kg/ha) is included in the seeds mixture. This distinction has been borne out by experience in many of the experiments with white clover carried out on our research stations.

There is still a need to have white clover varieties for the hill that are able to start growing earlier in the season than those commonly in use. It may be that it is the temperature requirement of the *Rhizobium* and the nitrogen fixation process rather than growth processes in the plant which limit earlier season growth. Work at HFRO by Vernon (1978) indicated that strains of *Rhizobium* differ in their ability to infect roots, form nodules and fix nitrogen at a temperature of 6-7°C so it may be possible to select or improve some of these so that they will complement any new low temperature tolerant varieties of white clover which the breeders may find.

Companion grasses

Some aspects of the comparison of grass species for improved hill pastures were examined in earlier work at HFRO (Hughes and Nicholson, 1961/2). These studies indicated that from a persistency point of view, and in the absence of data on quality, the following could have a place in some situations: red fescue (*Festuca rubra*), timothy (*Phleum pratense*), perennial ryegrass (*Lolium perenne*), cocksfoot (*Dactylis glomerata*) and Yorkshire fog (*Holcus lanatus*). Workers at the Scottish Plant Breeding Station have suggested that some species or inter-specific hybrids of *Poa* may also have a part to play.

It is theoretically attractive on both biological and economic grounds to retain the bulk of a piece of acid grassland (*Agrostis-Festuca* spp) while adding lime and phosphorus and introducing white clover. Unfortunately, this is still a relatively difficult procedure and more work is needed to devise reliable and rapid techniques of clover introduction and establishment. The technique of introducing white clover into the relatively less competitive swards of the *Nardus/Deschampsia/Agrostis-Festuca* type in wetter conditions has been shown to succeed but the quality of the grass which accompanies the clover is not good. Many studies have shown that herbage quality is of great importance to the nutrition of sheep (Chapter 4) and therefore replacement of the existing grasses with better quality species might be sound economic practice.

Considerable controversy centres around the species and varieties of grass which should be sown with white clover. Present HFRO practice is to sow ryegrass of several different heading dates, with timothy for wet sites and cocksfoot for dry sites. It is known that these species of grass require higher fertility, i.e. more nitrogen, phosphorous and potassium, than other species such as red fescue or Yorkshire fog which are sometimes recommended. This higher fertility is generated partly by the initial fertiliser addition, needed for the introduction of white clover, and partly by the use of controlled grazing and enhanced nutrient cycling (Chapter 5). Unfortunately, it is extremely laborious and expensive to carry out a full scientific evaluation of different grass species including the final assessment of impact on animal output and definitive data are lacking. Until the data are forthcoming the HFRO policy is to use varieties of lowland species which do well in the tests carried out, usually under lowland conditions, by the Agricultural Development and Advisory Service and the Scottish Colleges of Agriculture.

Use of herbicides

For grass

Dalapon was the first successful herbicide for grasses to be marketed in the United Kingdom. Recommendations for its use were based on trials carried out by several research organisations including HFRO. The trials which were carried out at Sourhope were described by Davies, Hunter and King (1960) and King and Davies (1963).

In summary, it was discovered that dalapon, when applied to an indigenous sward at relatively low rates (approximately 5.6 kg/ha) could act in a selective way and effectively control *Nardus stricta* and *Molinia caerulea* whilst leaving the more useful species such as *Agrostis* and *Festuca* unaffected.

Attempts to oversow seeds on to swards treated with dalapon gave variable results which were generally poor. The herbicide was applied at relatively high rates (11.2-22.4 kg/ha) to achieve maximum kill of the indigenous sward, followed by the addition of lime and phosphate and the broadcasting of grass and clover seed six weeks later. In addition, a pitch-pole harrow was used in some of the trials in an attempt to break the surface mat. Failures were attributed to several factors — the low rainfall of the south east of Scotland, the problems caused by the surface mat in the establishment of the sown seed, which were only partially overcome by the use of the harrow, and the resistance of certain species, such as *Holcus lanatus, Anthoxanthum odoratum* and *Poa pratensis,* to the spray even at high rates of application.

Paraquat which, in recent years, has replaced dalapon to some extent came on the market in 1961. Its advantages over dalapon were a more rapid uptake by the plant making it less affected by rain falling soon after application, together with its immediate absorption and inactivation in most soils. The latter advantage meant that grass and clover seed could be sown within a period of a few days of the herbicide spraying rather than the six weeks needed with dalapon.

Trials carried out by HFRO on paraquat (HFRO 3rd Report) were not as extensive as the ones using dalapon and were limited to a comparison

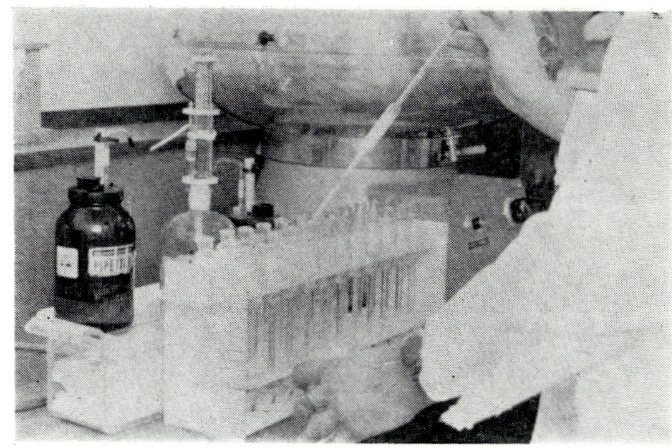

Milk analysis

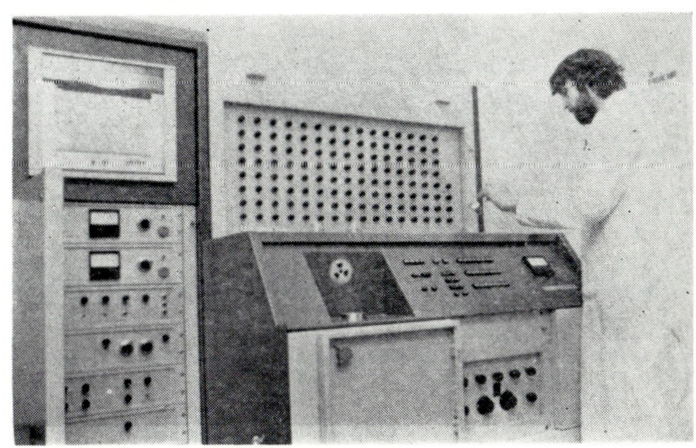

Inorganic
analysis

Soil phosphorus
analysis

Some analytical procedures at Headquarters

of the killing properties of the two chemicals. In general, there was little difference between the effect of applying either herbicide on the grass sward although there was some slight evidence that paraquat was more successful against *Festuca ovina* which had proved rather resistant to dalapon. However, the use of paraquat on matted herbage over peat or raw humus suggested that the herbicide was not broken down and deactivated as on mineral soils and some evidence of residual toxicity, particularly to white clover, was found. Such effects do not occur with dalapon.

For bracken

HFRO interest in the bracken (*Pteridium acquilinum*) problem can be traced back to the reports of the Scottish Hill Farm Research Committee which advocated the setting up of HFRO. Their Second Report (DOAS, 1953) concludes that 'the only practicable method so far available of controlling bracken was by destruction of the fronds at an appropriate stage of growth'; this approach was understandable, since at the time of reporting no suitable herbicide was available. Trials were set up in the Bowmont Valley involving the testing of bracken control machinery and, although results showed that eradication of bracken by mechanical means was generally effective and economic, a changing economic situation soon made it a less attractive technique. At a later stage HFRO co-operated with the Scottish Institute of Agricultural Engineering in studying the frequency of cutting and bruising necessary to prevent the substantial regeneration of bracken on areas in the Bowmont Valley, which had previously been brought under control in the earlier trials. Results, however, were inconclusive due to large seasonal variations which masked any possible treatment differences.

Experiments were carried out at Sourhope between 1959 and 1963 (HFRO 3rd Report) applying either Weedone Brackcontrol or Weedazol-tr (activated amino-triazole) but effects were short-lived. Two other herbicides (dicamba and pichloram) became available soon after the above trials were concluded but results were variable. It was concluded that more information was required on the effect of climate, soil type and stage of growth to explain these variations. It was not until Holroyd, Parker and Rowlands (1970) reported on asulam that it became evident that an effective herbicide giving consistent results had been found. Asulam has since been extensively used by farmers and is the only herbicide at present which receives a Government grant towards the cost of bracken eradication, provided its use is coupled with fertiliser treatment. Another herbicide currently under test is glyphosate which compares favourably with asulam in reducing frond numbers but is more destructive of the underlying vegetation.

Field trials were set up at Sourhope in 1973 to measure increase in herbage production following the use of asulam. These were confined to areas where the bracken density was such that the underlying vegetation, although perhaps sparse in places, was at least present and dominated by either *Agrostis* or *Festuca* species. The trials are continuing and results up to 1976 have been published (Davies, Newbould and Baillie, 1979). The main conclusions were that asulam, applied in July at the commercial rate of 4.5 kg/ha gave a 98% reduction in frond numbers which persisted for three years. A gradual increase in bracken then occurred but five years after spray application there was still a 90% reduction in fronds.

The use of asulam gave an increase in yield of grass of only 18% where botanical composition and soil conditions were adverse. Yield increases on better sites were improved to 50% and the additional grass yields resulting from bracken control amounted to approximately 1000 kg/ha. Because of the present high cost of the herbicide and its application it is doubtful if areas with potential increases in grass production less than 1000 kg/ha are worthy of treatment. The substantial increases in yield on the more inherently productive areas, however, make them more amenable to treatment. With a high degree of utilisation such improved areas could probably support an increased stocking of one ewe per hectare, particularly if used in conjunction with larger areas of indigenous vegetation as in the two-pasture system.

More information is required on whether the use of lime and phosphate is always justified with the spray treatment and this is currently under investigation at HFRO. There is also a lack of information on the most suitable techniques to be used in areas of dense and vigorous bracken with no underflora and where some form of reseeding operation is necessary to create a sward.

Seed bed preparation

Cultivation

No cultivation, other than treading by grazing animals is needed in wet 'open' (i.e. less competitive) sward situations. In the drier, more podsolised sites, however, it is often necessary to cultivate to open up the surface mat, expose soil, control indigenous vegetation and bury litter. Cultivations can range from harrowing or disking, through rotovating, to ploughing and even to sub-soiling should there be sub-surface pans in the soil.

It is not easy to be precise about cultivation procedures to use in individual situations. The nature of the 'ideal' seed bed required for arable crops has not yet been described for United Kingdom lowland conditions so it is no surprise that we are a long way from defining the precise soil conditions required to optimise the early growth of white clover and grasses in hill soils. Thus, problems occur which range from complete failure, through patchy and slow seedling establishment, to the rapid disappearance of some species and the speedy incursion of undesirable unsown species such as *Cerastium* (chickweed), *Carduus* and *Cirsium* (thistles), *Juncus* (rushes), *Nardus*, *Holcus* and *Bromus*. The consequence is a lack of impact on animal production, a waste of money and bad publicity for the role of land improvement. Causes of these difficulties at germination are lack of or too much water, low temperature, toxic chemicals, pests and diseases, too much litter, acidity and salt effects from locally applied fertilisers. Later, difficulties in early plant growth may be caused by lack of nutrients or uneven distribution of nutrients, competition from regenerating indigenous vegetation or germinating weeds, and too early defoliation by grazing animals.

Sowing date and germination

The time of sowing is probably the most critical factor which influences germination. The ground must be warm and moist during the period immediately following sowing. Late spring (May-early June) is often chosen to meet these requirements but, especially in the east of the country, a period of dry weather can occur from late June to August which can impair the further establishment and root growth of seedlings. Rogers and Bruce (unpublished) have shown that several species of grass benefit from late autumn sowing but white clover does not. Although seeds of the latter germinate more rapidly than grass after autumn sowing the seedlings are soon killed by frost. However, the presence of hard seeds (i.e. those unable to absorb water rapidly and to commence germination) in some varieties of white clover can result in an increase of seedlings the following spring. It would appear that two sowing times are needed to optimise seedling establishment of grass/white clover mixtures; as the effects of interspecific competition have not been investigated no firm recommendations for this policy can yet be made.

A further complication is caused by the application of compound fertiliser with the seeds and its influence or germination. Perennial ryegrass, in both a peat and a mineral soil, germinated more slowly when the starter fertiliser was applied with the seeds than it did when applied either 10 days before or 10 days after sowing. The overall situation was less clear with white clover but germination in peat was also slower when the fertiliser was applied with the seeds. Laboratory experiments indicated that the germination of white clover was particularly sensitive to all levels of potassium nitrate while that of ryegrass was only affected by the highest level (equivalent to 400 kg/ha of a compound fertiliser).

Some work to determine temperature thresholds for germination of different grass and white clover varieties was undertaken and considerable differences were found. For example, the lowest temperature for 50% germination to occur ranged from 2.9°C for S.22 Italian ryegrass (*Lolium multiflorum*) to >7.5°C for S.48 timothy; Huia white clover was intermediate at 4.9°C. Information of this type may help to determine the most appropriate seed mixtures, dates and season for sowing.

Investigations have been started to examine the suggestion that chemicals (described as allelopathic) exuded or left over from some species of the previous vegetation, interfere with germination in the same way as the chemical fertiliser described earlier. In preliminary experiments extracts from *Festuca rubra*, *Eriophorum vaginatum* and *Pteridium aquilinum* were found to reduce or stunt the root length of both white clover and perennial ryegrass seedlings while those from other hill species had little or no effect (Rogers and Bruce, unpublished).

Different methods have been investigated to introduce white clover into species-poor acid grassland on brown earth soil after adding lime and phosphate and without destroying all the grass vegetation and turf. No one method has been found satisfactory and it is disappointing that the idea of forcing seeds into the mat with compressed air proved of little value (Rogers, 1977).

Lime and fertilisers

The general requirements are closely related to those of white clover and there is great interest in determining the minimum applications needed to keep costs down. There has to be a balance between the achievements of sufficient biological response at an economic cost and maximum response at high levels of application.

This is particularly true of lime where the application to hill soil of procedures for determining requirements as used for lowland arable soils can suggest a need for as much as 50 tonne/ha to bring the top 15 cm of soil to pH 6.0. However, it has been shown that this amount is unnecessary. In peat soils it is only necessary to raise the surface pH to about 6.0 and in mineral soils to reduce the quantity of soluble aluminium, which interferes with plant growth, by raising the pH to at least 5.8 (Floate, 1978b). Liming to satisfy 50% of the base exchange capacity (BEC) is a good indication of the amount that will be adequate for most hill soils. The application of this principle to field conditions is currently being studied in experiments to assess the lime response of reseeded pastures on different soils.

With phosphorus, analysis by traditional methods is again difficult with hill soils (Floate and Pimplaskar, 1976) because of complications due to organic phosphorus and to soluble aluminium. A detailed comparison of a number of methods indicates that the quantity of total phosphorus is as good a guide as any (Pimplaskar, Floate and Newbould, unpublished). This follows because the supply of available phosphorus in hill soils is maintained by current mobilisation, predominantly from the organic form, which is the major component of the total phosphorus supply.

There has been little work on potassium until recently but it is now evident that on deep peat soils it is necessary to add sufficient potassium to match the quantity of available phosphorus and that the recycled potassium from grazing animals is rapidly lost from the soil in these situations (Floate *et al*, unpublished).

Response to addition of trace elements has been examined as part of other experiments but no consistent benefit has been achieved.

Nutrient requirements to maintain pasture

Little is currently known about the precise nutrient maintenance requirements for sustained herbage production from improved or reseeded pastures on hill soils. However, data from early laboratory studies on decomposition (Floate, 1970d), from current experiments on field litter decomposition and from the input/output experiments at Sourhope and Lephinmore (Chapter 5) are being used in the development and validation of a phosphorus cycling simulation model. This work is being done in collaboration with workers at the CSIRO Pastoral Research Laboratory, Armidale, Australia. It is anticipated that this work will assist in the long term interpretation of nutrient utilisation and recycling and lead to better predictions of the maintenance requirements for phosphorus on a range of soil and vegetation types in different environmental situations.

Some studies have already been carried out to assess maintenance requirements on deep peat at Lephinmore. Following the laboratory studies on the decomposition and mineralisation of nutrients from plant material and excreta, and concurrently with the development of improved systems of hill sheep management, a series of plot experiments on deep peat was set up to determine the nutrient requirements for establishment and maintenance of improved pasture, and to examine the role of nutrient recycling via grazing sheep. Because of the limited value of indigenous blanket bog vegetation (Chapter 2) all plots were reseeded with ryegrass and clover following establishment dressings of lime, nitrogen, phosphorus, potassium and trace elements. After the first season, when 2.5 tonne/ha lime only had been applied to grazed plots, the sown species were markedly reduced and production fell below 2000 kg DM/ha. Herbage production from treatments with 5 tonne/ha lime was only marginally higher (up to 2500 kg DM/ha). Low production levels and loss of sown species were not simply due to low pH but other reasons were not at first clear. Pot experiments (Rangeley and Newbould, unpublished) and an experiment designed to diagnose nutrient deficiency in the field all suggested that potassium deficiency aggravated by inadequate lime treatment, was the most likely cause.

In 1977 field-scale grazing trials, following the treatments of plots with potassium and phosphorus, produced total herbage yields in excess of 5000 kg DM/ha with both ryegrass and clover showing highly significant responses. This result was in contrast to cut-plot experiments where the only significant response to added potassium was from clover. The evidence suggests that the effect on grazed areas may be due to a direct response to applied fertiliser reinforced by an indirect response to recycled nitrogen. On phosphorus and potassium treated plots about 20 kg/ha of nitrogen and potassium were returned in a two week grazing period — about ten times the amounts measured on ungrazed, cut plots.

Thus, with added phosphorus and potassium, production levels on reseeded ryegrass/clover swards on deep peat can be double the amounts hitherto attained, but the maintenance fertiliser requirements to sustain these levels of production are not yet known. Experiments are currently in progress to determine the amount and frequency of fertiliser needs on this and other soil types.

Long term field experiments to assess more precisely the needs for lime have also been set up on several soil types in areas with different rainfall. It has been a generally accepted practice since the early sixties (Hughes and Nicholson, 1961) to apply 2.5-5.0 tonne lime/ha every 5-7 years to maintain a pH between 5.5 and 6.0, the amount and frequency of the addition depending primarily on rainfall. In the same situations phosphorus at 40-60 kg/ha as ground mineral phosphate would normally be applied every 3-4 years. This latter is particularly important to maintain a good proportion (20-30%) of white clover in the sward. The need for potassium on deep peat soils has already been described, but on mineral soils there appears to be an adequate return of potassium from the grazing animal which is then retained by the soil.

Policies of applying fertiliser for maintenance will help to keep both white clover and sown grasses more sensitive to soil fertility in the hill sward. Any relaxation of this policy is quickly followed by a disappearance of white clover and the increased reappearance of indigenous grasses such as *Festuca, Agrostis* and, in some cases, *Nardus.*

Responses to pasture improvement

Many factors influence the responses in pasture production and quality to improvement by the procedures outlined here, and the only true test is to measure the impact of the improved pasture on the output of animal product and on farm income (Chapter 6). However, the summarised results given in Table 4 attempt to show likely levels of response in yield and quality based on HFRO field data for the main soil/vegetation groups as shown in Table 1. Utilisation of indigenous pastures in the traditional system varies from 5-20% and in all these pastures, except acid grassland, increases in utilisation brought about by control of grazing have little effect on yield or quality. However, an increase in utilisation from 20% to 30%

Table 4. Likely responses of four main indigenous hill vegetation types to pasture improvement by moderate control of grazing or to establishment of sown pasture with good control of grazing. Estimated average annual levels of dry matter yield (kg/ha) and seasonal range of digestibility (DDM-%)

| | Indigenous pasture | | | | Sown pasture | |
| Sward type* | Little grazing control | | Moderate grazing control | | Good grazing control | |
	Yield	DDM	Yield	DDM	Yield	DDM
Acid grassland (*Agrostis-Festuca*) spp. poor	2500	76-40	2800	76-50	6000	78-66
Dry shrub heath	2000	60-40	2000	60-50	5000	78-66
Wet grass heath	1500	70-35	1600	72-55	4500	78-66
Bog	1400	68-40	1400	68-40	4000	78-66

* See Tables 1 and 2 for detailed description.

of the acid grassland may raise yield by 10% and quality by 2-3 units of digestibility. The real impact on herbage yield and quality therefore comes from a combination of control of grazing and pasture improvement. Levels of utilisation of up to 70% are possible on sown ryegrass/white clover pasture. Grazing of high quality herbage at this level not only has a considerable impact on animal performance (Chapter 4) but is also beneficial to the continued well-being of the pasture. The data also indicate that for all except the dry shrub heath vegetation *(Calluna)* the quality of the herbage (% digestible dry matter) can be relatively high at least for short periods of time in the spring.

The future

Work in the next few years will aim to define more precisely the responses in herbage production to be expected from the range of hill soil types according to the quantities of lime, phosphate and the other treatments it is possible to afford within the economic limitations of the farming system. In addition, it is hoped to define or refine procedures which will remove much of the present unpredictability of establishment and performance which can accompany land improvement in the hills, particularly as regards the growth of white clover. The consequences of level and severity of utilisation by sheep and/or cattle at the seasons which suit the strategies of the two-pasture system will also need to be assessed from the point of view of defining 'ideal' sward structure and compostitions (species and varieties) and understanding the recycling of nutrients through the grazing animals which should permit more precise definition of the levels of fertiliser need for maintenance. It is anticipated that more work of a similar nature to that described here for hill soils will be needed with upland soils and pastures. Dependence on the legume for provision of nitrogen may be less in this situation, and the use of nitrogen fertiliser to lengthen growing season in spring and autumn and to even out dips in production at mid-season may need to be quantified.

References

DOAS (1953). *Hill Farm Research Second Report.* The Scottish Hill Farm Research Committee. Edinburgh, HMSO.
HOLROYD, J, PARKER, C and ROWLANDS, A (1970). Asulam for the control of bracken *(Pteridium aquilinum* (L) Kuhn). *Proc. 10th Br. Weed Control Conf.* 1, 371-376.

4

Hill sheep production and nutrition

Contributors: Rhd H Armstrong, J M Doney, R G Gunn, J A Milne, J N Peart,
       A Whitelaw

Co-ordinators: A J F Russel and J Eadie

Introduction

A major part of the Organisation's research effort in the last twenty-five years has been directed towards the many facets of the production from and nutrition of hill sheep. Production and nutrition in any form of livestock enterprise are inextricably interwoven, and to put the Organisation's contribution in these areas into its proper context it has to be viewed against the background of the traditional system of hill sheep management described in Chapter 1; it is against that background that this chapter considers some of the main areas of work carried out in the Organisation during the last twenty-five years on hill sheep production and nutrition.

Nutrition and the components of production

The Organisation's earliest reports indicate that the importance of nutrition in hill sheep production was appreciated from the beginning. Although work on lactation, lamb fattening and wool growth was initiated within the first few years, the main emphasis of the nutritional research at that time was concerned with the problems of winter nutrition. The HFRO Second Report, for example, reviews the results of 'new investigations directed primarily towards the problems of the "hungry gap" period from January to April, when hill pastures provide only a minimum of growth and when the majority of hill ewes, though pregnant, live on a sub-maintenance diet and decline in weight'. It was not until the early to mid-sixties, however, that a systematic examination of the relationships between nutrition and the many factors contributing to production from hill sheep — reproduction, pregnancy, lactation, lamb growth and wool growth — was begun.

Nutrition and reproductive performance

The high potential reproductive capability of the hill ewe, and particularly the Scottish Blackface, has long been recognised. The early work of the Organisation on the supplementary feeding of hill ewes during pregnancy indicated that, even under traditional systems of grazing management, reproductive performance could be influenced by nutrition. It was observed, for example, that ewes fed during the winter 'might be expected to recuperate better in autumn so making for improved performance in the following year. These deferred advantages are in evidence at Glensaugh where the pre-mating weight of ewes and the incidence of twinning in different sub-flocks vary with the liberality of pre-lambing feeding' (HFRO 2nd Report).

The inadequacy of the nutritional cycle afforded by the traditional system of management is responsible in a number of ways for the moderate or low levels of reproductive performance. In many situations it prolongs the period of growth and development and is associated with very low levels of twinning in the younger age groups (Gunn, 1967c). Although the incidence of twinning increases with age from around 5% up to about 40%, the overall flock response is severely depressed by the low levels in the younger but numerically-larger ages. Improved rearing of replacement ewes, however, does little to raise the level of adult reproductive performance on the hill, and only when such improvements are matched by comparable improvements in adult nutrition is there a worthwhile response (Gunn, 1972).

The removal of the inadequacies of nutrition at times of year other than winter, by the provision of high levels of nutrition throughout summer, before and during mating, and during late pregnancy and early lactation, produced dramatic responses in reproductive performance and confirmed that the reproductive potential of most hill breeds is severely depressed by nutrition in adult life under hill conditions. There was thus a clear need for more detailed studies of the nutritional factors involved in this depression (Gunn, 1967c).

The research programme which started in 1967 attempted to quantify the patterns of energy intake during those different stages of the annual cycle which influence the stimulation of oestrous activity within the normal breeding season, the shedding of ova by the ovaries, the fertilisation and survival of these ova and the maintenance of the resultant embryos to term as viable lambs. Most of the work has been carried out on the Scottish Blackface breed, but a number of studies on North and South Country Cheviots, and to a lesser extent on other breeds and crosses, has indicated some genotypic differences in response, if not in principle at least in degree (Doney and Gunn, 1973; Gunn and Doney, 1974; Gunn, Doney and Smith, 1979a, b). It is therefore not yet possible to extrapolate and draw up a general hypothesis to cover all breeds.

Ovulation rate. The nutritional aspects affecting ovulation rate which have received most attention are body condition and the level of nutrition in the pre-mating period. The effect of body condition on the number of ova shed is dramatic. Early work on Scottish Blackface ewes established that, within a flock, the mean ovulation rate of those individuals brought into a good body condition at mating (condition score 3) could be in excess of 200% or a mean of two ova per ewe with little variation between individuals. This compared with a mean ovulation rate of little more than 100% when ewes were in poor condition (condition score 1.5) (Gunn, Doney and Russel, 1969). Subsequent work has shown that when ewes are in intermediate levels of body condition the mean ovulation rate is also intermediate (Gunn and Doney, 1975; Doney and Gunn, 1979). The relationship between body condition at mating and ovulation rate has been found to exist, to a lesser extent, in other breeds and crosses, although the range of ovulation rate varies with genotype and standard of life-time nutritional management. In all breeds where ewes are below a certain low level of body condition oestrus and ovulation are suppressed. As condition increases above this level so does the likelihood of more ova being shed per oestrus, and consequently on a flock basis the ovulation rate increases. A knowledge of the body condition of individual ewes does not, however, allow predictions to be made of how many ova each ewe will shed (Russel, Doney and Eadie, 1978). Once the maximum genetic ovulation potential of a flock has been attained further increases in body condition appear to have no positive effect and may even lead to reductions in ovulation rate.

Body condition at mating is largely a function of previous nutrition during the recovery period between one reproductive cycle and the next. The pattern of change in condition during recovery does not itself affect ovulation rate (Gunn and Doney, 1973b). Body condition can therefore be regarded as a medium-term factor affecting ovulation rate, as distinct from growth which is a long-term factor, and level of nutrition during mating which is short term.

The level of nutrition in the immediate pre-mating and mating period has much less of a direct effect on ovulation rate in Blackface ewes than does the level of body condition (Gunn *et al*, 1969; Gunn and Doney, 1975). Only within a relatively narrow band of body condition does current nutrition have an effect additional to that of achieved body condition. Above and below this band it is condition itself which is important and not the current level of intake. In breeds other than the Scottish Blackface this interaction between body condition and current nutrition may be of greater significance and effective over a different, and perhaps wider, range of body condition (Gunn *et al*, 1979b).

A further series of studies has demonstrated that, irrespective of the level of nutrition, ovulation rate may be significantly depressed by exposure to climatic or management stresses, particularly if those are imposed in the period up to one week before mating (Griffiths, Gunn and Doney, 1970; Doney and Gunn, 1972; Doney, Gunn and Griffiths, 1973; Doney, Gunn, Smith and Carr, 1976). In one experiment, in which the ewes were in a uniform moderately-good condition, the imposition of premating stress depressed the ovulation rate from 186% to 152%. Similar orders of depression have been found in ewes in different body conditions and subjected to various stress factors.

Embryonic mortality. Ovulation rate, however, is only half the story. The ova, having been shed, have then to be fertilised, and the embryos implanted and carried through to parturition. While ovulation can now be regulated with some precision by nutritional management, embryo wastage does not, so far, lend itself to such precise control. Values for embryonic mortality ranging from 15-50% have been recorded in different situations. A similar range of nutritional and non-nutritional factors as influence ovulation rate also affect embryo wastage (Gunn and Doney, 1975; Doney, Smith and Gunn, 1976), but the recorded variation implies a range of interactions, as yet not fully understood. A programme of study to elucidate this is under way.

These differences in ovulation rate and embryo wastage lead to considerable variation in achieved lambing percentage. This can range in hill flocks from less than 80% under extensive conditions to 160% or more in nutritionally improved situations.

The objectives of reproductive studies in the hill sheep breeds are the quantitative establishment of optimum growth and development patterns and within-season nutritional and management standards for the control of reproductive response in all environments.

Nutrition during pregnancy

Early studies. The nutrition of hill ewes during pregnancy and particularly in the weeks immediately before lambing, was one of the first research topics to be identified at the time of the Organisation's inception. The second Hill Farm Research Committee had reported that 'the hill sheep occupies a unique position among farm livestock in that it is required to survive during winter on a level of nutrition which is frequently insufficient to maintain body-weight, whilst its annual pregnancies coincide with this very period of semi-starvation' (HFRO 1st Report).

Field-scale feeding trials with concentrates, turnips and hay, conducted principally by J F Robinson, were started in the winter of 1955/56. In the subsequent few years approximately half of the 2500 ewes on the Organisation's research stations received some form of supplementary feeding for six to eight weeks before lambing, the remaining ewes acting as controls by which the effects of the supplements could be judged. The conclusions drawn from this early work were that at 'Glensaugh and Lephinmore, where ewes wintered on the hill without supplements would produce lamb crops between 60% and 70%, the effect of supplementary feeding has been to raise the lamb crops by more than sufficient to cover the cost. On the other hand, at Sourhope where a normal lamb crop is between 90% and 100% the effect of supplementation has been too slight to justify it as a regular practice under existing conceptions of management' (HFRO 2nd Report).

The published account of further studies (Robinson, Currie and Peart, 1961) underlined the advantages to be gained from supplementary feeding in late pregnancy, not only in the poorer nutritional environments of Glensaugh and Lephinmore, but also in the better grassy hill situation typified by Sourhope. The benefits of improved pre-lambing nutrition were recognised as affecting not only lamb birth weight, but also ewe fitness and, in turn, ease of parturition, maternal instinct, lactation performance and the subsequent recovery of live weight and condition. Many of the husbandry recommendations made as a result of these early studies are still being urged on the industry today.

The large-scale pre-lambing feeding trials were concluded after six years. It was recognised at the outset, however, that 'the elucidation of the more fundamental aspects' of late pregnancy nutrition would require 'closer experimentation with smaller groups of sheep' (HFRO 1st Report). This more detailed second stage of the work began in the early sixties and in some respects continues today. The starting point of this phase of the work, which received a considerable impetus when Dr R L Reid was appointed Associate Director, was the assessment of the nutritional status of hill ewes during late pregnancy.

Assessment of nutritional status. The term 'nutritional status' is used to describe the adequacy, or more often the inadequacy, of the ewe's food intake in relation to its nutrient requirements. During pregnancy hill ewes are invariably undernourished, for example their energy intakes are insufficient to meet the requirements for energy to maintain the ewe herself and to support the maintenance and growth of the foetus and other products of conception. In such situations the ewe mobilises her own body tissues — principally fat and protein. The extent to which body fat is being mobilised, and hence a measure of the severity of undernourishment, can be assessed from the concentration of certain metabolites in the blood.

One of the principal metabolites first used in this way was plasma free or non-esterified fatty acids (FFA or NEFA) (Russel, Doney and Reid, 1967a). Plasma FFA concentration is a very sensitive index of moderate degrees of undernourishment, but is less satisfactory in situations of very severe or prolonged undernutrition. The concentrations of ketone bodies (β- or 3-hydroxybutyrate (3-OHB) and acetoacetate) on the other hand, show only relatively small increases in moderately undernourished animals, but are markedly elevated in severe undernourishment, being most responsive at levels where FFA concentrations become less useful. A combination of plasma FFA and ketone concentrations can therefore be used to monitor with precision the whole spectrum of undernourishment. Plasma FFA concentrations are, however, very susceptible to certain hormonal changes which follow any upset or disturbance of the animal, and are consequently of limited value under field conditions with ewes which are unaccustomed to frequent handling or to the procedures associated with blood sampling. Plasma ketone concentrations are less readily affected by such factors and are generally a more reliable index of nutritional state in any work with large numbers of animals under practical conditions (Russel, Maxwell and Foot, 1973).

The biochemical method for determining total ketone bodies in blood samples does not lend itself readily to the automation necessary in large-scale experimental work with flocks of pregnant ewes where there is often a need for more or less immediate measurements of nutritional state. In an attempt to overcome this difficulty the separate use of either of the main ketone bodies was examined. Plasma 3-OHB concentration is relatively easily measured using modern laboratory technology, but because a certain quantity of this metabolite is derived from dietary sources through rumen butyric acid, it is not always immediately apparent how much of the measured concentration stems from mobilised body fat. In theory, plasma acetoacetate concentration should provide a more easily interpreted index of the rate of fat catabolism, but its biochemical determination requires a degree of sample handling and preparation which is rarely possible when working under field conditions at some distance from the laboratory. For these reasons plasma 3-OHB concentration is now the preferred and most frequently used index of energy status (Russel, 1978b).

Late pregnancy. The early work on pregnancy nutrition attempted to relate production at lambing time directly to the quantity of supplements fed in late pregnancy. The ability to assess the energy status of the ewe added the missing factor of energy deficit required to balance the equation which has energy intake from grazed herbage plus supplementary feeding on one side, and energy requirements for the ewe and the products of conception on the other.

Preliminary studies on ewes from the Low End flock at Lephinmore showed that the severity of undernourishment in late pregnancy, as assessed by plasma FFA concentrations, was closely related to foetal weight. From this and other work at Sourhope it became clear, firstly, that the relative degree of undernourishment of a group of ewes was determined by their level of nutrition and, secondly, and perhaps of greater importance, that within a particular nutritional spectrum the absolute severity of undernourishment in individual animals was determined by foetal weight.

This understanding of the relative importance of the main factors affecting nutritional state in the pregnant ewe led to a realisation of the inadequacies of conventional experimental designs. It became clear, for example, that in an experiment with two or three fixed levels of feeding the variation in nutritional state within animals on any one level of feeding was likely to be greater than the mean differences in feed intake between groups. Using FFA and ketone concentrations as a basis to adjust individual feed intakes to maintain prescribed nutritional states, work was carried out which provided basic and much needed information on energy requirements of the hill ewe in late pregnancy, and on the effects of varying degrees of undernourishment on factors such as lamb birth weight, ewe milk production and early lamb growth rate (Russel *et al*, 1967a, b; Peart, 1967a).

Foetal energy requirements in the final stage of pregnancy were established to be 1.5 MJ ME/kg foetus/d, a figure which has subsequently

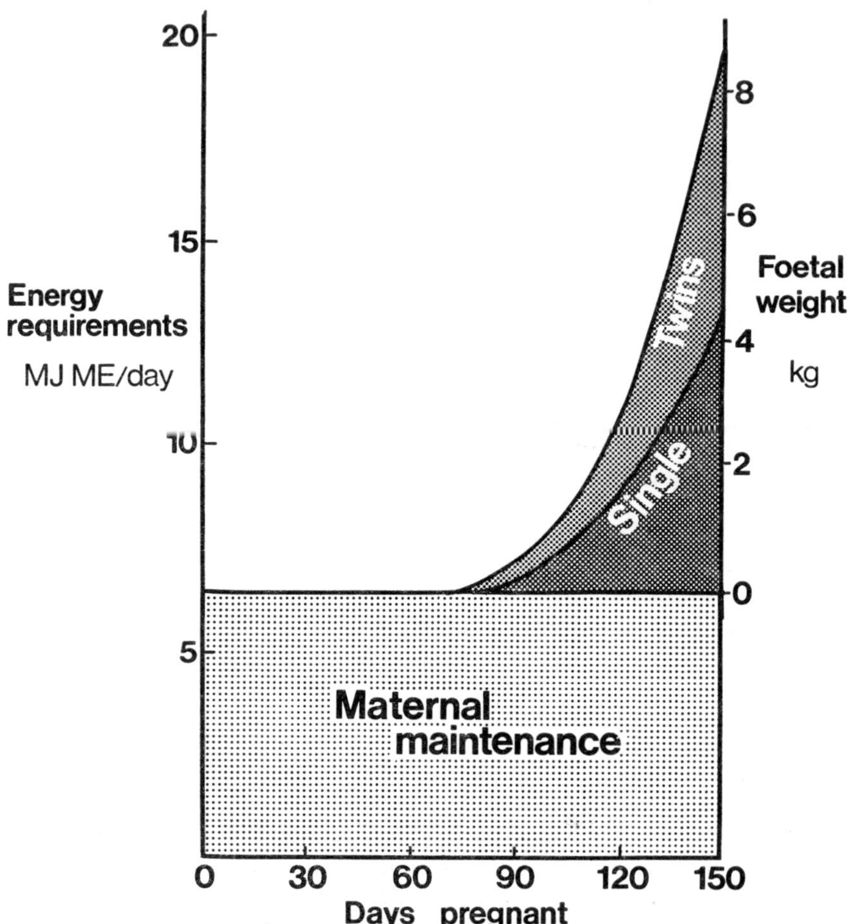

Fig. 1   Theoretical amounts of energy required to prevent undernourishment during pregnancy in housed 50 kg ewes carrying single or twin foetuses.

been confirmed by other workers. This estimate was much higher than had been commonly supposed at that time, and meant that immediately before lambing a 50 kg ewe with a 4.5 kg single lamb would have an energy requirement approximately twice that of a non-pregnant ewe, while a ewe with twins would have a requirement about two-and-a-half to three times that of a non-pregnant ewe (Fig. 1). These very high values, however, represented full requirements in the strict sense of the term, and were not recommended levels of feeding. Indeed the same work showed that energy intakes which were some 25% lower than full requirements would reduce

Controlled feeding of ewes & lambs

Hand milking of ewes

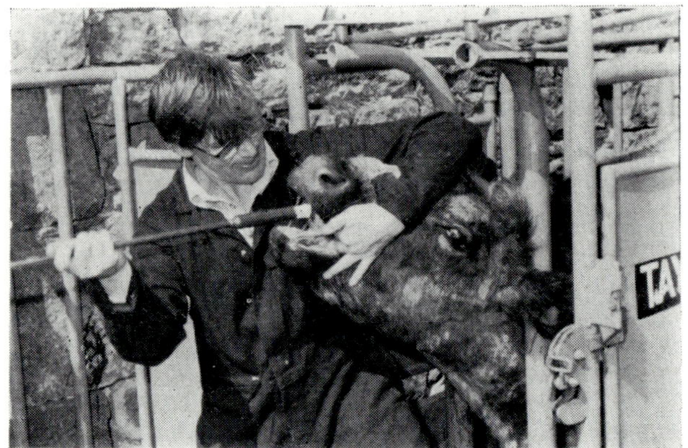

Chromic oxide dosing

lamb birth weights by less than 10% — a production penalty which was and is still regarded as acceptable on both biological and economic considerations. The shape of the lamb birth weight/energy intake response curve indicated, however, that a further lowering of the level of intake would quickly lead to reductions in lamb birth weight which were not acceptable and where lamb survival and the subsequent growth rate of those which did survive was prejudiced.

This and later work relating biochemical measurements of nutritional state to energy balance led to the development of the concept of what constituted an acceptable nutritional state in late pregnancy. This is a state in which the ewe is fed less than full requirements and is consequently catabolising body tissue to meet the energy deficit, but which, at the same time, ensures satisfactory weight and vigour of the new-born lamb.

The more recent work on late pregnancy nutrition has been largely concerned with the application of the earlier findings from closely controlled experimentation with individually-fed, housed ewes to the flock situation. In dealing with flocks of ewes comprised of individuals varying in intake, stage of pregnancy and foetal numbers, the achievement of the same mean energy status as befits a single individually-fed ewe would result in a significant proportion of animals being very severely undernourished. The ewes experiencing the greater nutritional penalty would invariably be those carrying twins, i.e. the ewes in which it is most important to minimise the production penalty. Some compromise is therefore necessary, and this can only take the form of maintaining a less severe degree of undernourisment for the flock as a whole. Present evidence suggests that this means lowering the plasma 3-OHB threshold at which supplementary feeding is given from the value of about 1.1 mmol/l, which would be appropriate for individually-fed ewes, to about 0.8 mmol/l.

Some of the practical recommendations which are currently advocated for late pregnancy supplementary feeding of hill ewes, such as the more generous nutritional treatment of gimmers and lean ewes, have remained unchanged for the last 15-20 years, although their scientific basis is now better understood. To these have been added the recommendations of identifying and feeding separately early and late lambing ewes, and of matching the pattern of feeding to that of foetal growth to make most efficient use of the ever more costly concentrates. Although the regulation of feed inputs on the basis of analyses of blood samples, which is now standard practice in the Organisation's development projects, is unlikely to have application in commercial hill sheep enterprises, the research work based on this approach has allowed patterns of weight change, as an indication of the adequacy of feeding, to be defined.

Perhaps the greatest single limitation to the improved nutritional management of hill flocks in late pregnancy is our inability to identify those ewes carrying twins. Various attempts have been made in the Organisation over the last fifteen years to find a technique which would enable foetal numbers to be determined speedily, accurately and at a reasonable cost. Concentrations in the blood of various metabolites, enzymes and hormones have been examined, and although some of the techniques have yielded promising results, it is considered that this approach is unlikely to be practicable in commercial enterprises. Radiographic and ultrasonic techniques can work well and be justified in experimental situations, but for a variety of reasons are not well suited to widespread use on hill farms in their present forms. The problem is, however, still judged to be of major importance and work in this area is continuing with some hope and expectation of eventual success.

Mid-pregnancy. Nutrition during mid-pregnancy, i.e. during the second and third months, has generally been thought to be of little consequence, particularly when considered in relation to the importance of late pregnancy nutrition. Some of J F Robinson's early work on supplementary feeding included treatments which 'indicated that sub-standard nutrition in mid-pregnancy detracts from the efficiency of response to extra feeding in the last month of pregnancy' (HFRO 2nd Report), but little was done in this field until the seventies.

Investigations into the cause of poor birth weights of gimmers' lambs showed that the problem was not due to inadequate late pregnancy nutrition, as had first been thought. Examination of the interaction between the level of nutrition during mid-pregnancy and the weight, size and condition of gimmers at first mating showed that with small, poorly-grown gimmers in only moderate condition at mating, a nutritional regime which prevented loss of maternal weight during mid-pregnancy resulted in higher lamb birth weights than a regime in which the animals lost some 5 kg of maternal weight over the same period. Conversely, in well-grown gimmers in good condition at mating the same higher level of nutrition caused an appreciable reduction in birth weight compared with the lower plane (Russel, Foot and White, unpublished). The underlying physiological reasons for these apparently contradictory results have since been elucidated by work at the Rowett Research Institute.

Apart from the obvious implications of improving the nutrition of poorly-grown gimmers and of ensuring some weight loss during mid-pregnancy in well-grown gimmers, these results suggest that improving the nutrition of ewe lambs and hoggs during the summer months, through higher quality pasture, would ensure a heavier weight and better body condition at first mating.

There is evidence to suggest that older ewes are better buffered against the effects of both over and undernutrition than are gimmers, but it is likely that the results of the work on pregnant gimmers will also apply in some measure to mature ewes. At about the time the results from the work on gimmers became available, it was considered that the role of mid-pregnancy nutrition in the adult ewe should be reappraised. A number of studies in the development programme had by then reached the stage where the increased number of ewes being carried on the fewer and poorer acres of the unimproved hill grazings during the winter months was causing considerably greater losses in ewe live weight and condition during mid-pregnancy. Late pregnancy nutrition was by this time closely controlled, and there was no direct evidence of any adverse effect on production of the presumed lower level of mid-pregnancy nutrition, but it was considered pertinent to examine the effect of nutrition during mid-pregnancy, and perhaps of more importance, the effects of its interaction with nutrition at other stages of the annual cycle, on production. This work is currently in progress at Lephinmore, and it would be premature to discuss the results in any detail at this stage. Suffice to say that, while there are interesting indications of effects of mid-pregnancy nutrition on lambing percentage and lamb birth weight, both affecting the total output of weaned lamb, these appear to be of lesser consequence than the benefits to be gained by better summer nutrition through the utilisation of improved pasture.

Body composition in pregnancy. The importance of use of the ewe's body reserves to meet nutritional deficits was appreciated at an early stage, and at about the same time as the more detailed pregnancy/nutrition studies began, work was started, again at Lephinmore, to quantify the changes occurring in the body composition of hill ewes during pregnancy. Earlier work in the late fifties had shown that the loss of live weight over winter was considerable, often being of the order of 8-9 kg. The body composition studies showed, firstly, that the typical hill ewe at peak condition in the autumn carried little enough fat as judged by normal standards — about 6 kg or 12-14% of live weight — and secondly, that more than half of this very modest reserve was catabolised during pregnancy, leaving little in the way of reserves for lactation. It was shown, too, that some 20% of the ewe's body protein and minerals were used during this time. Studies of individual fat depots indicated that more than 85% of subcutaneous fatty tissue had been used by lambing time and that considerable amounts of fat were withdrawn from the skeleton in the last month of pregnancy (Russel, Gunn and Doney, 1968). Although this information on compositional changes throughout pregnancy was to prove very useful, the main value of the work perhaps lay in highlighting the paucity of the traditionally-managed ewe's energy reserves at the beginning of winter, and the extremely low level of body fat reached at the end of pregnancy, which does not, of course, signify the end of the period of dependence on body reserves. The work on changes

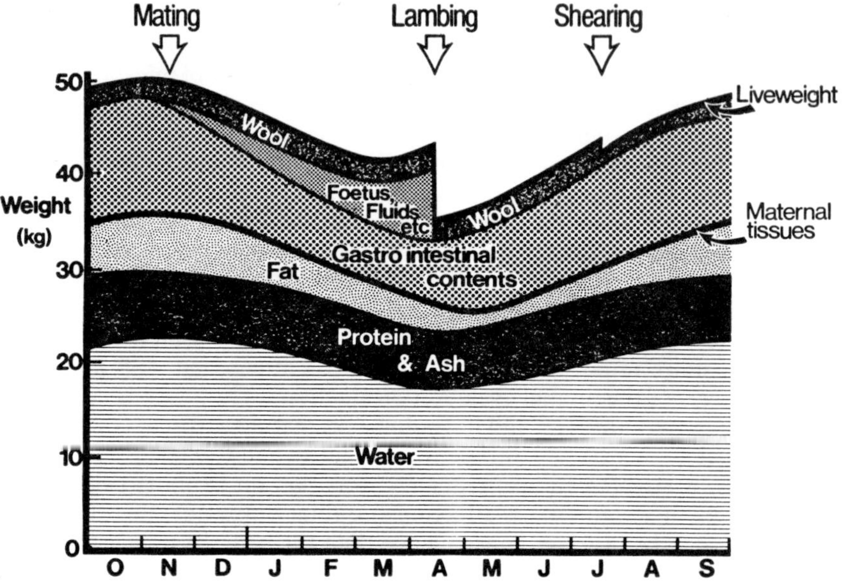

Fig. 2   Stylised illustration of changes in ewe live weight and body composition
throughout the year and in a poor nutritional environment.

in body composition and the mobilisation of body reserves still constitutes
a major part of the Organisation's research programme. The stylised
patterns of change in ewe live weight and in the weights of body fat and
protein throughout the year (Fig. 2) have been prepared from recent studies
with Blackface ewes at Lephinmore (Russel, Foot and McFarlane,
unpublished).

Changing attitudes. The early work on the supplementary feeding of
pregnant hill ewes was carried out against a background of scepticism and
predictions that any form of feeding, other than perhaps a little hay when
the ground was completely covered by snow, would destroy the hardiness,
the raking ability and all other inherent virtues of the Blackface ewe. It was
clearly the prevailing opinion that no good would come from this work;
attitudes, however, have changed. Twenty-five years ago the few flock-
masters who did practice some feeding were loathe to admit it and reluctant
to discuss it. Today some of the questions most frequently posed by hill
sheep farmers are concerned with quantities and protein levels of pre-
lambing supplements, and the discussion centres on whether the ewes do
better on cobs or blocks. It is not possible to gauge the effect which the
Organisation's research work in this field has had in shaping the opinions

held and the feeding strategies practised today, but the effects of the changed attitudes on the industry as a whole have been considerable.

Nutrition in relation to lactation and lamb growth

The conditions required to produce satisfactory lamb crops have already been discussed. Subsequent production is dependent on mortality and the growth rate of surviving lambs. Survival of new-born lambs is influenced by their own vigour at birth and by the maternal care and onset of lactation by the ewe. These factors are dependent on the nutritional state of the ewe and on the general environmental conditions at parturition.

On many hill pastures, where the quality of the herage is already relatively low and begins to decline by the end of May (Eadie, 1967b), lamb growth rate progressively declines as milk yield falls. Typical growth patterns for several hill breeds and their crosses have been established in a range of grazing and supplementary feeding conditions (for example Munro, 1962a; Doney and Munro, 1962; Peart, 1967a; Peart, Edwards and Donaldson, 1972; Peart, Doney and Macdonald, 1975). The results have demonstrated that hill lambs have a growth potential much greater than normally achieved in farm practice. The wide variation in growth rate found in these studies, though partly dependent on genotype, is mainly determined by the milk and solid food provided for the lamb throughout the pre-weaning period. The work has been carried out and is conveniently discussed under two related headings: the factors which directly affect milk production by the ewe, and the factors which influence total nutrient intake and utilisation by the lamb.

Milk production. The series of lactation studies has established the levels of milk production and the shape of lactation curves of several British hill breeds, strains and their crosses in both pen-fed and grazing ewes with restricted or unrestricted access to food. In general terms, the factors which are responsible for variation in the quantity and quality of milk production are now known and can be used to enhance lamb production in commercial practice.

Contemporary ewe nutrition is the most important single factor determining milk production, particularly in the early stages of lactation. It has been established that the yield of well-nourished hill ewes suckling twins rises rapidly to a maximum in excess of 3 kg/d by two to four weeks after parturition, then steadily declines to around 1 kg/d by the twelfth week. Single-suckled ewes attain peak yields of approximately 2 kg/d by three to five weeks and decline to levels similar to the twin-suckled ewes. This pattern of milk production and the characteristic shape of the lactation

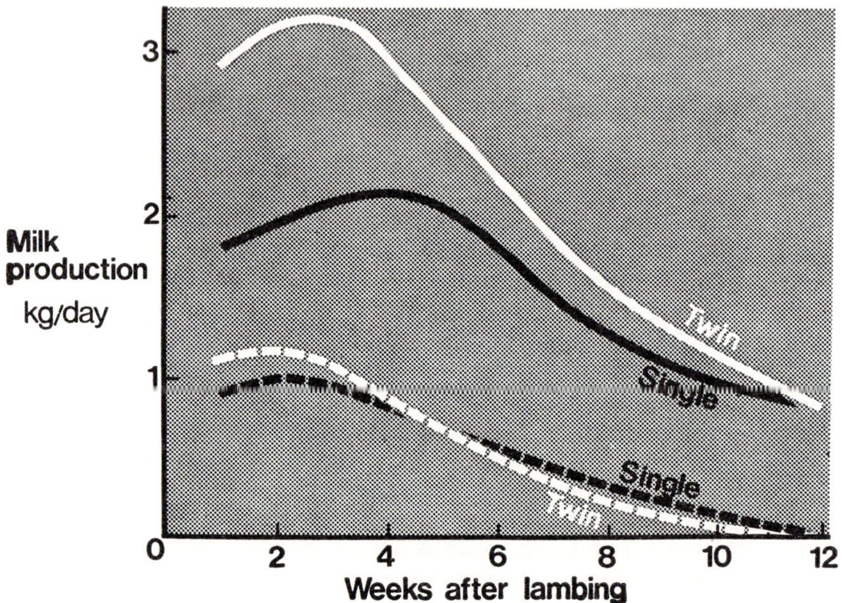

Fig. 3 Milk production of single- and twin-suckling ewes grazing sown (solid lines) and poor quality unimproved (broken lines) pasture.

curve, which is illustrated in Fig. 3, is common to all the native breeds and crosses studied. Under hill grazing conditions, however, these potential levels of production are rarely achieved (Peart, 1970a) because of the limitations of hill pasture as a source of adequate nutrition for the ewe. The extent of the depression depends on characteristics of the pasture (Eadie, 1967b; Milne and Peart, unpublished). For example, and as shown in Fig. 3, when grazed on low quality *Molinia*-dominant pasture the yields of both twin- and single-suckled Blackface ewes reached a peak of only 1 kg/d and declined to less than 300 g/d by eight weeks (Doney, unpublished).

The nutrient requirements to achieve the inherent potential levels of milk yield are much higher than at any other stage of the production cycle, and those for multi-suckled ewes are obviously greater than for ewes with single lambs. Food intakes of ewes on concentrate diets increase rapidly after parturition, attaining maximum values around five weeks before steadily declining (Peart, 1967b), but the rate of expansion and total grazing intake may be considerably modified by the characteristics of the available herbage (Eadie, 1967b; Maxwell, Doney, Milne, Peart, Russel, Sibbald and MacDonald, 1979; Foot and Tissier, 1977; Doney, unpublished). The expansion of appetite of twin-suckled ewes is normally greater than that of single-suckled ewes, although their capacity to increase intake may be limited if food intake has been severely restricted before parturition or if they are offered a diet of low digestibility (Peart, 1967b; Maxwell *et al*, 1979; Foot and Russel, 1979; Doney, unpublished).

Despite the possible effect on appetite, the influence of pregnancy nutrition has been found to have only a marginal effect on subsequent milk production of Blackface ewes when fed on high quality pelleted food *ad libitum* (Peart, 1967a) or grazed on high quality pasture (Eadie, 1967b; Doney, unpublished) during lactation.

Reserves of body condition at parturition act as a buffer between nutrient intake and nutrient requirement of ewes during early lactation. This factor is of considerable importance in hill flocks where the grazing, especially in early lactation, may be inadequate for satisfactory milk production. However, this ability to utilise body reserves in the absence of adequate current nutrition cannot maintain the full potential yields. Results have indicated that the rate of conversion of body reserves is more critical than the absolute amount of reserves available during this period although, if restricted feeding is continued beyond the third week, ewes in better body condition can maintain greater milk yields than can lean ewes. It has also been found that the milk yield of previously under-nourished Blackface ewes increases in response to increase in available nutrition in early lactation, whereas, if the improvement in nutrition is delayed to the fourth week or later (after the time at which peak yield is normally reached), there is no corresponding increase in milk production (Peart, 1968b, 1970b).

Altough nutrition during lactation is the dominant factor affecting milk production, other factors have some direct influence. The effect of number of lambs suckled has already been described. Most of this difference is found in the first six weeks. Peart *et al* (1972) found that milk yield ranked in order with the number of lambs suckled during the first three weeks, after which the yields rapidly converged. However, the practice of early removal of one member from a twin pair does not allow the remaining twin to grow at a greater rate than a normal single lamb (Doney and Munro, 1962).

The quantity of milk extracted from similar ewes may be influenced by the genotype of the lambs, for example Texel x Blackface lambs suckled 23% more milk from pure Blackface ewes in the whole lactation than did pure-bred Blackface lambs (Peart *et al*, 1975). Similarly, the nature of the alternative solid food available to lambs can influence the pattern of lactation of the ewe (Doney and Peart, 1976).

Finally, milk production may be influenced by the genotype of the ewe. Within breeds, considerable variation in peak yields and rate of decline can be found amongst ewes given equal nutritional opportunities.

Conversely, the mean yields and shape of lactation curves do not differ significantly amongst the native breeds studied when ewes are maintained in equivalent conditions. However, the use of crosses between native breeds and breeds selected for dairy production can result in higher peak yields, and especially in a significantly more sustained pattern of lactation (Louda and Doney, 1976; Krizek, Louda, Jakubec and Doney, 1978; Doney, Peart, Smith and Louda, 1979; Peart, Doney and Smith, 1979). These advantages were found to remain when crossbred ewes from a dairy sire on Blackface and pure Blackface ewes were grazed on *Agrostis-Festuca* or on a sown upland pasture (Peart and Doney, unpublished).

The problems concerning relationships amongst voluntary intake, pasture characteristics and the partition of intake between milk or body tissues in different breeds or crosses and in different nutritional environments require further study to develop a general concept of planned management, but the findings can already be used effectively to improve performance in specific systems.

Lamb growth. It has been recognised that pre-weaning growth rates of lambs reared under hill conditions are normally well below their genetic

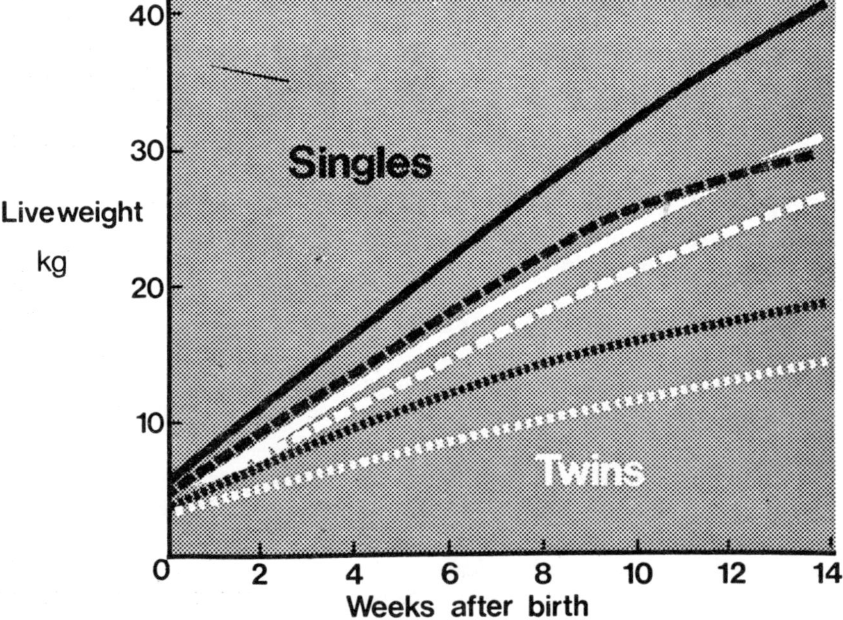

Fig. 4   Live-weight (solid lines) changes of single and twin lambs suckling ewes grazing sown pasture (broken lines), high quality unimproved *Agrostis-Festuca* pasture (broken lines) and low quality *Molinia*-dominant pasture (dotted lines).

potential and that this depression of growth rate is most marked after the first few weeks of lactation. For example, mean growth rates in intensively-reared, pure-bred Blackface single lambs have been recorded around 350 g/d in the first four weeks of lactation, rising to over 400 g/d between the fourth and twelfth weeks. In contrast, mean growth rate over several years in the hill flocks was around 230 g/d from birth to marking and this declined to less than 160 g/d between marking and weaning (Armstrong and Eadie, 1973b). These variations in growth rate are illustrated in Fig. 4.

During the first few weeks the lamb is entirely dependent on milk, and variation in growth rate is almost completely determined by the factors which influence milk yield and quality. It has been shown that lamb factors may themselves modify milk yield in that, for example, small undernourished lambs at birth may be unable to utilise all the available milk, and this in turn leads to a depression in milk production. Conversely, a vigorous demand for milk by the lambs in the early stage may stimulate a higher, sustained level of production by ewes (Peart et al, 1975).

From the age of about four weeks the lamb begins to consume an increasingly significant amount of solid food and the interactions of the mixed diet become more complex. Preliminary studies illustrate this complexity. The availability of milk, as affected by ewe nutrition, by the number of lambs suckling each ewe, or by lamb genotype differences in the suckling drive, can influence both the age at which lambs begin to consume significant amounts of solid food and the rate of expansion of their intake (Peart, 1967b; Peart, 1970a). This interaction between two progressively changing components of diet is further modified by certain characteristics, especially digestibility, of the solid food available to the lamb (Armstrong and Eadie, 1973b; Doney and Peart, 1976).

It has been suggested that the characteristics of pasture ingested by the lamb in the second half of lactation are probably more responsible for the decline in growth rate of lambs grazed on hill pastures than is the level of ewe nutrition in this period. Milk yield is already declining rapidly at this time and would not be sustained in the native breeds by improved nutrition. Hence lamb growth rates in late lactation, although depending partly on their early growth rate, which in turn depends on their milk intake, are largely determined by the amount and quality of herbage ingested. The provision of improved pasture during the lactation period results in significantly increased lamb growth rate, both through increased levels of digested herbage and increased milk availability (Armstrong and Eadie, 1973b, 1977; Doney, unpublished). As an alternative approach it has recently been shown that the use of ewes with an enhanced potential for sustained lactation can support a higher growth rate of lambs in a range of nutritional conditions including concentrate feeding, upland pasture and hill pasture (Peart et al, 1979; Doney, Peart and Smith, unpublished).

The complex relationships amongst the consumption of milk and herbage by lambs and the utilisation of the mixed diet remain the most important single problem. Preliminary work indoors and at pasture using lambs weaned from their mothers but fed controlled amounts of milk substitute suggested that the advantages of higher quality herbage for lamb growth were still apparent with this 'milk' supply. This was so because depression of herbage intake due to milk was slight and largely independent of herbage quality (Armstrong and Eadie, 1973b).

After weaning the lamb is entirely dependent on intake of solid food. The importance of herbage quality and the limitations of hill grass species for lamb growth have been demonstrated by Armstrong and Eadie (1973b, 1977). Improved management of the better quality native hill grasses such as *Agrostis-Festuca* may result in some increased lamb growth, but this is small in comparison with the known potential. A larger increase in growth can be achieved on sown grass pastures. However, at the time when pasture becomes the main source of lamb nutrients, the available herbage is declining in digestibility. The intrinsic characteristics of white clover are particularly valuable as its digestibility declines more slowly than does that of the grasses. Furthermore, at the same level of digestibility, voluntary intake of clover tends to be higher than that of grass and the efficiency of utilisation of digested intake for growth also appears to be higher (Armstrong and Eadie, 1973b). These advantages are most important in a situation where lambs with high appetite drive and growth potential are otherwise dependent on herbage of inherently low quality.

More recently, studies on the identification of other management factors and food sources which influence the variation in pre- and post-weaning growth and body composition have been initiated (Foot, Doney and Maxwell, unpublished; Peart, Foot and Doney, unpublished).

Ewe recovery. The importance of a high and sustained milk supply for growth of lambs cannot be over-emphasised. Nevertheless, the consequences to the ewe of a high level of lactation cannot be ignored; it must be considered as part of the whole integrated sequence of reproduction and lamb growth. The role of body condition at mating has already been described. Studies on the response of milk production to nutritional variation would be incomplete without some consideration of the effect of this production on the long-term recovery of body reserves which are likely to be already depleted at lambing. Changes in live weight during the course of lactation have been studied in several of the nutritional situations reported. The results confirm the generally accepted concept that, except where ewes are fed *ad libitum* on a high quality pelleted diet or are on high quality grazing, live weight continues to fall during the immediate post-partum period and thereafter recovers steadily throughout lactation. The duration and extent of the fall in live weight and the subsequent rate of recovery have been found to be related to initial body condition, to the rate of milk production and to the quantity and quality of the available food resources. It is apparent that in the lactating ewe the partition of digested nutrients between maintenance and production is a complex process dependent on the integrated hormonal state of the animal as well as on inherent genetic factors. Definitive studies on intake partition are complicated by the limited value of change in live weight as an indicator of change in body composition over this critical, short period. In recently initiated studies more precise information on body composition changes during lactation is being obtained. This will allow a better understanding of the factors which influence intake and the partition of nutrients, especially in early lactation where the requirements for established milk production are likely to exceed the current levels of nutrient intake. It is noteworthy, however, that general indications given in a number of experimental situations suggest that even when minimal recovery has been achieved by weaning time, the period between then and mating could still be sufficient to establish the required body condition in ewes, provided that nutrition is adequate.

Nutrition and wool production

In the early years of the Organisation a number of studies on the factors affecting wool production were carried out. The objectives of the programme were to describe the growth pattern of the fleece of hill ewes under standard hill management, to identify the factors which are associated with variation in growth rate and fleece quality, to determine the relative importance of these factors in terms of fleece weight and value and, finally, to interpret the results in terms of their application to economic performance in new systems of management.

The first observational studies on Blackface sheep showed that the variation in wool growth rate of hill-grazing ewes throughout the year was one of the main limitations to annual fleece weight (Doney and Smith, 1961a, b). The hill breeds were found to be similar to unimproved breeds, such as the Soay, in that the rate of fibre growth in winter may be as low as 10-20% of the maximum growth rate found in the summer months. This contrasts with sheep selected for wool production, such as the Merino breeds, which may show only a small seasonal amplitude in growth.

The relative contributions of variation in the qualitative components, such as fibre density, length and diameter, to total fleece weight of the Blackface were assessed (Doney, 1963b). It was concluded that quality so defined was largely unrelated to fleece weight, and that variation in the latter was the significant factor in the economic return from wool. It was also found that nutrition during embryonic and early growth stages had only a marginal effect on adult wool production (Doney and Smith, 1964). Specific aspects of fleece or fibre quality were reported in relation to the occurrence of abnormal fibres (Smith, unpublished) and the casting of whole or parts of the fleece before shearing (Doney and Smith, 1969). Further studies on this latter and important factor have failed to identify any simple management procedure which would prevent its occurrence.

Differences in growth patterns between strains of Scottish Blackface and other hill breeds, such as the Cheviot, were described (Peart and Ryder, 1954; Doney and Smith, 1966) but these were found to be small in terms of overall return on wool production.

Whilst the association between wool growth rate and seasonal variation in food intake of grazing hill sheep was noted (Doney and Eadie, 1967) it was established that the marked annual rhythm found in the hill breeds was determined by inherent regulation and that, unlike the Merino, low growth rates during winter could not be improved by better nutrition (Doney, 1966b, 1967a, b). In the hill breeds, however, there was a very significant increase in fleece weight associated with improved nutrition during lactation and during the post lactation period in summer and autumn (Fig. 5). Variation in season, nutrition and physiological state were found to affect the chemical composition, especially the sulphur content of the wool (Doney and Evans, 1968) but this has little relevance to fleece value.

Since output of wool is a relatively small component of total flock returns (some 15-20% of income, including subsidies, depending on both reproduction rate and the relative cash values of wool and lamb) it was concluded from this series of studies that improvements in management should be directed entirely at increasing the economic efficiency of production of weaned lamb (Russel, Doney and Maxwell, 1976). Such management improvements can be expected to, and indeed have been found to, carry with them some significant increase in amount of wool produced on a flock basis. Conversely, a management system designed to maximise wool production per head would carry few benefits in terms of increased lamb production.

Production of wool as a saleable product, however, is not the only consideration. The fleece, especially of hill sheep, can be regarded as a protection against excessive heat loss and, therefore, a contributor to reducing the energy cost of maintenance in a cold, wet and windy environment (Doney, 1963a). General studies have shown that adaptation of sheep to the hill environment depends on factors other than the fleece, for example sheltering behaviour and skin-temperature regulation (Munro, 1961; Griffiths, 1967; Doney and Russel, 1969). The reduction of wool

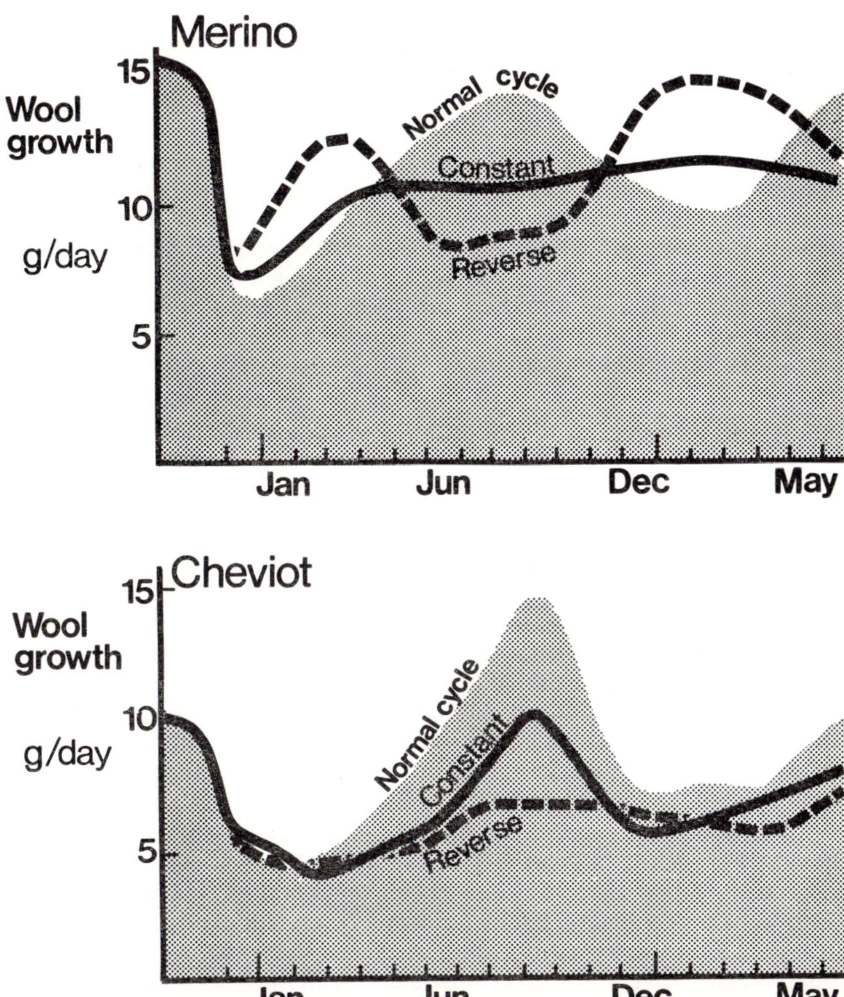

Fig. 5   Wool growth in Merino wethers and Cheviot wethers on a constant level of nutrition throughout the year; on a regime following the normal annual nutritional cycle; and on a reversed annual nutritional cycle.

growth found in winter may even be partly dependent on the other physiological adjustments to cold; Doney and Griffiths (1967) found a reduction in wool growth associated with a lowering of skin temperature. The general conclusion from these studies was that the possession of good fleece cover by the beginning of winter is an essential condition for the general well-being of extensively-grazed hill sheep, but that variations either in fleece type or in growth rate during winter were relatively unimportant. Hence, as was the case with nutrition, it can be suggested that no particular effort need be expended either on genetic change of fleece type or management designed specifically to change wool production.

As a consequence of this the research effort directed to wool has been considerably reduced. Changes in both quality and quantity associated with changes in management systems and with changes in breed are continuously monitored. The growth characteristics and manufacturing properties of fleeces from Texel × Blackface have been compared with those of Border Leicester × Blackface and pure Blackface reared in the same environment. This work was carried out in co-operation with the Wool Industries Research Association (Smith, unpublished).

Year-round nutrition

The results of the programme of research on the effects of nutrition on the many components of production demonstrated very clearly the interaction and inter-dependence of the various phases of the annual cycle. It became even more apparent that hill sheep production must be considered on a year-round basis, and that any action in one part of the cycle will have a reaction in some other part. Perhaps the best example of this is the calamitous situation which could arise from improving nutrition before mating, resulting in higher ovulation and conception rates, without making provision for the additional nutritional needs in late pregnancy and early lactation.

The synthesis which followed this analytical stage also indicated a change of emphasis, at least as far as individual animal performance was concerned. Whereas there was a preoccupation in the Organisation's early years with the problems of winter nutrition, by the late sixties the benefits to be gained from improved nutrition at every other time of the year were apparent. Improved nutrition from better quality pasture had been shown to be particularly important in the spring for higher levels of lactation and early lamb growth, in the summer to sustain these higher lamb growth rates and to allow some live-weight recovery in the ewes, and in the autumn to bring the ewes into the better body condition required for higher ovulation rates. With this enhanced condition at mating the ewes went into the winter with more reserves on which to draw, and the earlier problems of nutrition throughout the long winter months appeared to recede, or at least were seen in a different perspective.

The account of the systems studies which appears in Chapter 6 provides a clear picture of the very considerable responses in individual animal performance which were achieved in the early seventies by the use of improved pasture in the context of the two-pasture system. The various development projects were not, however, concerned only with output per ewe. As ewe numbers were increased over the years towards and beyond the point at which output of weaned lamb per ewe was reduced — and as experimental projects they had to be taken to these limits — so it became more difficult to attain the higher pre-mating live weights and levels of body condition achieved in the initial stages. At the same time the rates of loss of live weight and condition over winter increased, and it became apparent that further work on the nutrition of hill ewes during the winter months, and particularly over mid-pregnancy was required.

The tracing of the Organisation's work on the nutrition of the hill ewe provides examples not only of how certain problems or questions tend to recur, albeit in a somewhat different context each time, but also of how the research and development aspects of the work interact. The synthesis of the results of research on the effects of nutrition on the components of production led directly to the creation of new systems of management which, when rigorously examined in large-scale development projects, have identified areas, such as mid-pregnancy nutrition, requiring more detailed research work.

Specific production and nutrition topics

The supplementation of grazed herbage

In the Organisation's early work on supplementation of ewes in pregnancy attention was given to the type of supplement. At Glensaugh experiments were conducted in 1956-58 in which turnips, hay and concentrates were compared as supplements in the last six weeks of pregnancy of ewes grazing a heather hill (HFRO 2nd Report). Small differences between the effect of supplements on ewe and lamb performance were attributed to differences in estimated energy intake. No differences could be detected in the performance of ewes given concentrates ranging from 7% to 17% crude protein. It was appreciated at the time that the numbers of ewes used in the experiments and the lack of adequate measurement of roughage intake made it difficult to demonstrate statistically significant differences between supplements, or to interpret adequately the results obtained. A more valuable contribution made at that time was observations on the feeding habits of ewes given concentrates. It was found that concentrate feeding interfered little with the foraging ability of sheep and that, by feeding ewes only every second day, distances travelled by sheep could be reduced and competition between ewes for concentrates could be decreased to the potential advantage of the flock.

For six years between 1961 and 1967 no research was conducted on type of supplement although more precise information was gained on the energy requirements of the pregnant ewe and on the effects of under-nourishment in late pregnancy. In 1967 research of a more detailed nature was initiated to eludicate some of the principles involved in the provision of supplements. Initially experiments on the effect of the cereal source of the supplement on roughage intake and on digestion of a range of roughages of different quality were carried out on wethers (Lamb and Eadie, 1979). It was found that the amount of concentrates given and the nitrogen content of the roughages were the major determinants of roughage intake.

These experiments were conducted with dried roughages, i.e. hays and straws. It was felt that sufficient had been learnt from the use of these roughages by the early seventies and that studies using winter-quality *Agrostis-Festuca* and heather could be undertaken. Some intake and digesti-bility measurements of *Agrostis-Festuca* had already been made by Armstrong and Eadie (1977). Measurements of the intake and digestibility characteristics of heather showed that the apparent overall digestion of nitrogen was poor and that its availability in the rumen was low; this was related to the protein-complexing properties of tannins found in heather (Milne, 1974).

The experiments on both *Agrostis-Festuca* and heather were carried out with frozen herbage, the use of which was validated in other studies (MacRae, Campbell and Eadie, 1975). It was found that, unlike the situation with higher quality material, poor winter-quality herbages cut from indigenous pastures did not change markedly in their chemical com-position when frozen and thawed. It was therefore considered reasonable to assume that freezing would not subsequently alter the processes of digestion when such herbages were fed to ruminants. Work was also undertaken to confirm the normality of surgically-prepared animals (MacRae and Wilson, 1977) and on the development of an adequate dual-phase-marker technique with which to measure rates of passage of solid and liquid material in the gut (Evans, MacRae and Wilson, 1977).

More detailed studies on the digestion and nitrogen metabolism of unsupplemented *Agrostis-Festuca* and heather diets were then conducted (MacRae, Milne, Wilson and Spence, 1979). These showed that sub-stantially more non-ammonia nitrogen arrived at the duodenum of sheep on the *Agrostis-Festuca* diet than was consumed, and that the net availability of this non-ammonia nitrogen to the animal was 60%. The digestion of nitrogen in the *Agrostis-Festuca* diet was similar to that of poor quality hay diets. On heather diets there was also a substantial net addition of non-ammonia nitrogen anterior to the duodenum, but in this case the net avail-ability of this nitrogen to the animal was only 40%. This low availability was associated with the irreversible complexing of protein with the tannins present in the heather. The increase in non-ammonia nitrogen anterior to the duodenum with both diets was found to consist mainly of non-urea nitrogen secreted into the upper region of the digestive tract, rather than from the recycling of urea nitrogen via the rumen. An important implication of this finding lies in the potentially serious consequences to the nitrogen economy of animals on a predominantly heather diet.

With the *Agrostis-Festuca* diet used in these experiments there was no evidence that digestion was impaired by a nitrogen deficiency in the rumen. This was confirmed when supplements of 100-600 g rolled barley were given to wethers ingesting *Agrostis-Festuca* in a pen-feeding experiment and at pasture. Intakes of roughage declined by 0.5 g/g supplement given and were not altered by increasing the amount of nitrogen in the supplement. This indicates that when a cereal-based supplement is given to ewes grazing *Agrostis-Festuca* herbage in late pregnancy the supplement acts as a partial substitute for roughage.

Previous results with heather indicated that it might be possible to increase its intake by supplementation in such a way that the added nutrients were not acting as a substitute but as a true supplement. Small amounts of nitrogen (as urea) and energy (as starch) increased both the roughage and the total digestible energy intakes considerably (Milne, Christie and Russel, 1979). The effects of the total intake of soluble nitrogen, i.e. from the basal roughage and the supplement, on voluntary

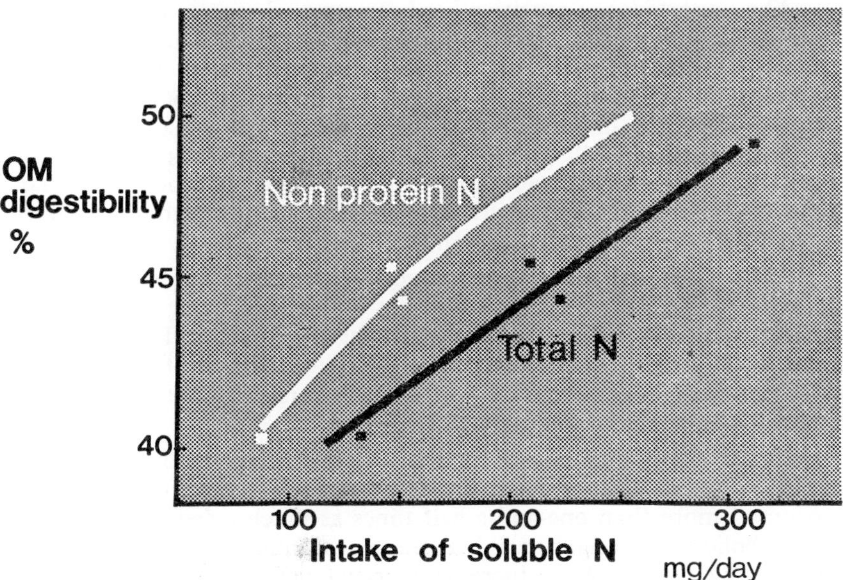

Fig. 6   The effect of soluble non-protein and total nitrogen intake on the voluntary intake and digestibility of organic matter from heather by sheep.

organic matter intake and on organic matter digestibility of heather, are illustrated in Fig. 6. The main use of these findings is in the mid-pregnancy feeding of ewes on heather-dominant hills where increased stocking rates can lead to unacceptably low levels of mid-pregnancy nutrition. Experiments on the feeding of supplements have been extended to examine the effects of these supplements on roughage diets containing proportions of *Agrostis-Festuca* and heather, and it has been shown that up to one-third of the diet can be as *Agrostis-Festuca* without the supplements acting as a substitute for roughage intake. At present, this work is examining in more detail the digestion of these supplemented diets of *Agrostis-Festuca* and heather and the extent to which the protein digested is used as an energy source.

It is intended that this work will specify the type of supplement to be fed in different situations. However, the more practical aspects of supplementary feeding, such as the competition for supplements in group-fed animals (Foot and Russel, 1973; Foot, Russel, Maxwell and Morris, 1973) have not been neglected and will become a more important component of the research in the future.

Voluntary food intake

The importance of nutrition has been stressed repeatedly throughout this chapter and in other parts of this report. To improve nutrition, i.e. to increase the intake of energy or other nutrients by the sheep, it is imperative to have some understanding of the relative importance of those factors which control the amount of food which sheep will consume. This is particularly true in hill sheep production which involves poor-quality, bulky diets with low concentrations of nutrients, and where, at certain times, the animals are physically incapable of eating enough to obtain the nutrients required even to maintain live weight.

Early work at Sourhope, reviewed in more detail in Chapter 5, had shown how the annual cycle of nutrient intake by grazing wether sheep varied throughout the year with changing herbage quality (Eadie, 1967b). Later work in the seventies sought to examine and quantify some of the main factors of animal origin, as well as those pertaining to the diet, which determine an animal's voluntary food intake.

In a preliminary study of the effect of fatness on voluntary intake it was shown that over a relatively short period (six weeks) thin ewes consumed more than one-and-a-half times as much dried grass per unit of metabolic body weight as did fat sheep; voluntary intake decreased by some 3 g/kg$^{0.73}$ for each 1% increase in body fat (Foot, 1972).

Some of the earlier work on pregnancy nutrition (Russel *et al*, 1967a) had drawn attention to the unwarranted assumptions of equality of food intake which had on occasion been made in experiments conducted with group-fed ewes. The variation in intake between individual animals offered restricted amounts of food on a group basis was examined in a series of experiments. The results showed the importance of the physical form of the diet in determining, through its effect on rate of consumption, the extent of the variation in intake between individual animals. On a diet of hay supplemented with a small quantity of concentrates the differences in energy intake between animals in the upper and lower quartiles was 35%; this increased to 60% on a diet of 3.7:1 pelleted:chopped dried grass, and to more than 80% on a wholly pelleted concentrate diet (Foot and Russel, 1973).

With housed, pregnant ewes on a typical feeding regime supplying a mean of 0.75 kg hay/hd/d and an allowance of concentrates increasing with advancing pregnancy from about 100-450 g/hd/d, variation in total digestible dry matter intake was as high as 24%. Individual concentrate intakes were most variable where the group allowance of concentrates was small (Foot *et al*, 1973). Variation in energy status during late pregnancy in group-fed, housed ewes, calculated by comparing individually calculated requirements with measured energy intakes, is illustrated in Fig. 7.

In later large-scale experiments attempts were made to distinguish between short- and long-term effects of animal and dietary factors on voluntary food intake. In addition to providing valuable quantitative information on these factors in animals in different physiological states, these studies demonstrated the complexity of the problem and highlighted some of the many interacting factors which make it so difficult even to describe the pattern of intake over any period of time greater than a few weeks. On a single diet of constant composition intake is likely to be affected by a changed environment (usual at the beginning of most experiments), by seasonal effects (probably operating through changes in day-length), and by the effect of previous intake and diet quality (through changes in live weight and body composition).

Initially, and after any change in diet quality, intake was shown to be dependent to a major extent on apparent dry matter digestibility, but over longer periods of time (in this case up to 33 weeks) this well known and much quoted effect was negligible. Indeed, in some instances there was an apparent negative relationship between intake and digestibility, as those sheep on the higher quality diets became fatter and heavier and so reduced their intakes (Foot and Russel, 1978).

There are many reports in the literature of reductions in voluntary food intake during late pregnancy, but our studies suggest that, at least in some instances, this may merely reflect the typical pattern of an initial increase in intake seen at the beginning of most experiments, and which is followed by a decrease after a period of some six to eight weeks. On a poor quality

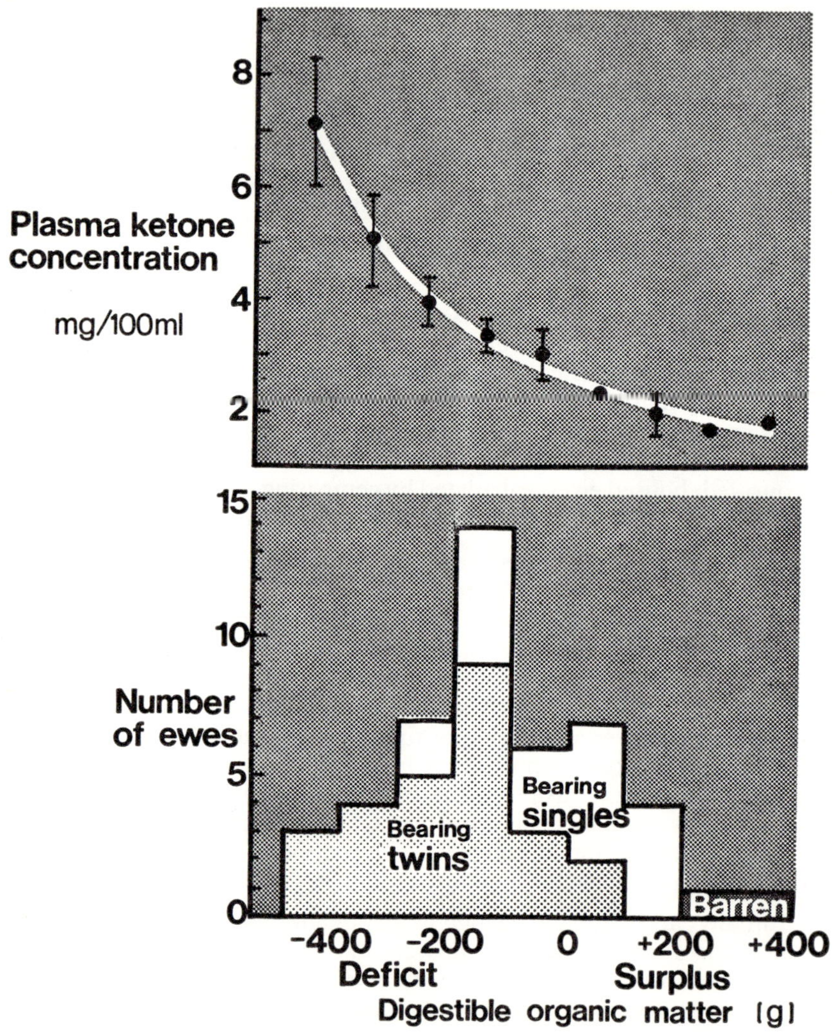

Fig. 7   Energy deficits and surpluses derived from measurements of food intake and calculated requirements and the relationship between these nutritional states and plasma ketone concentration.

roughage (51% digestibility) voluntary intakes were maintained at a relatively constant level throughout pregnancy. A decline in intake was, however, noted in late pregnancy in twin-bearing ewes consuming a high quality diet (70% digestibility), but at parturition their intakes were still some 12%

higher than they had been at the start of the experiment some 14 weeks previously. Overall there was no effect of number of foetuses on intake.

In lactation, ewes suckling twin lambs had higher voluntary intakes than ewes with single lambs. These studies support the view that the higher the diet quality the earlier and higher the peak intake. An effect of nutrition during pregnancy was also noted on voluntary food intake during lactation. The higher intakes during lactation of ewes which had been offered a low quality roughage in pregnancy were ascribed to their lower levels of body fat. Despite the number of variables measured in these experiments, the interactions between the many animal and dietary factors were such that no more than 64% of the variation in intake during lactation could be related to factors prevailing before and at parturition (pregnancy diet, ewe live weight and fat content) and during lactation (lamb live-weight gain and ewe live-weight change) (Foot and Russel, 1979).

As yet we do not have a comprehensive knowledge of the quantitative effects of the many factors and the ways in which they interact to determine voluntary food intake in the long term. These studies have, however, provided a valuable insight into many of the more important factors, and particularly those operating in the short term; they also provide a basis on which other work in related fields, such as the supplementation of grazed herbage, may proceed.

Mineral nutrition

Work on the mineral nutrition of hill sheep has been a feature of the nutritional research programme since the Organisation's inception. The major elements which have received attention include calcium, phosphorus, magnesium and sulphur, while the list of trace elements includes cobalt, copper, molybdenum and selenium. Most of the work carried out in the earlier years, before the Orgaisation had laboratories equipped to undertake the necessary analyses, was conducted in collaboration with other institutes, notably the Rowett Research Institute and the Animal Diseases Research Association. Although the Organisation now has its own laboratory facilities and can undertake at least the major part of the analytical work required for the programme on mineral nutrition, the links with the other institutes in this field are maintained, and the continuing discussions of proposals and results with these bodies are much valued.

Major elements. The early interest in mineral nutrition arose from the need to study the problem of premature wear and loss of incisor teeth in ewes — the so-called broken-mouth condition. A survey of this condition amongst ewe flocks in the north east of Scotland was conducted by Jones

(1958b) in the early fifties. This implicated the winter feeding of turnips as one of the predisposing factors and also drew attention to deficiencies of lime, phosphate and other elements in the soils of the area. Although this survey showed that the presence of abnormal wear in teeth appeared to be consistent with the hypothesis that soil or geological factors were the primary cause of the condition, it was also noted that the feeding of minerals was 'without discernible effect in alleviating the deterioration in the teeth' (HFRO 1st Report). Later, collaborative work with the Rowett Research Institute, which undertook radiological studies of broken-mouthed ewes, demonstrated severe erosion of the incisor alveolar bone of these sheep, but indicated that the skeletons were well mineralised and that generalised calcium and phosphorus deficiencies did not play an important role in the premature shedding of permament incisor teeth.

Other work from the Rowett Research Institute suggested, however, that skeletal repair in hill sheep was generally poor. An experiment at Glensaugh showed that calcium and phosphorus lost from the body during lactation on reseeded pastures were not fully replaced during the autumn recovery period on hill pastures, resulting in a gradual depletion of these minerals throughout the ewe's productive life, and possibly leading to demineralisation of the alveolar bone and premature loss of the incisor teeth (Gunn, 1969a).

Subsequent work on calcium and phosphorus metabolism in ewes at Glensaugh was carried out in collaboration with the Animal Diseases Research Association. This showed that both calcium and phosphorus contents of the skeleton fell between November and January, and again during lactation between April and July, although there was a marked increase in late pregnancy when supplementary feeding was given (Field, Suttle and Gunn, 1968). The losses of calcium from the undernourished pregnant ewe were judged to be abnormally high and it was considered that this loss may have been increased either by a deficiency of dietary protein or by an excessive catabolism of body protein to provide energy. Later work indicated that the winter protein deficiency probably resulted from the low digestibility of crude protein and drew attention to the likelihood of a phosphorus deficiency (Field, Sykes and Gunn, 1974). Other work about this time indicated that factors such as parasitism and trace element deficiencies could adversely affect voluntary intake of herbage, and thereby reduce energy and crude protein intakes, with consequent deleterious effects to calcium and phosphorus metabolism. Recent studies have confirmed the very poor protein status of pregnant ewes on the predominantly heather grazings at Glensaugh and Lephinmore (Sykes and Russel, 1979) compared to the apparently satisfactory status of ewes on the grassy vegetation at Sourhope.

Calcium metabolism also featured in collaborative work with the Glasgow Veterinary School on the effect of grazing management on clinical grass tetany. Plasma concentrations of calcium and magnesium were shown to decline rapidly when ewes were first transferred to the experimental plots, but clinical tetany developed only in the presence of both hypocalcaemia and hypomagnesaemia; it did not occur in the high proportion of ewes which showed low plasma magnesium values but which maintained normal plasma calcium concentrations (Hemingway, Ritchie, Brown and Peart, 1965).

Trace elements. Selenium has been investigated several times in collaborative studies with other bodies. The interest in this essential element probably stems from the dramatic responses noted on certain soil types in New Zealand, and to an extensive survey in the United Kingdom which gave reason to suppose that some response might be found in areas derived from Old Red Sandstone. It was considered, however, that responses were likely to be small, and work carried out with the Rowett Research Institute at Glensaugh was generally inconclusive. Later work at Glensaugh in the seventies failed to elicit any credible response to selenium injections.

Cobalt pine was the earliest, and one of the most important trace element deficiencies investigated in the Organisation. This was confirmed in a trial initiated at Sourhope in 1957 in which large responses to doses of cobalt sulphate were recorded in ewe live weights and in lamb birth and weaning weights. More recently, studies were undertaken at Sourhope to evaluate different means of prophylaxis. These showed that although cobalt bullets can on occasion be lost by regurgitation or rendered ineffective by encrustation, this means of therapy is generally most effective in producing rapid increases in serum vitamin B_{12} concentrations and in maintaining these at a high level in juvenile and adult sheep for a period of some years. Oral dosing with cobalt salts and the injection of vitamin B_{12} produced only transitory increases in serum vitamin B_{12} concentrations and had to be repeated frequently to be effective (Whitelaw and Russel, 1979a).

Cobalt deficiency has also been found to be of importance at Glensaugh where it was identified as the cause of ill-thrift in weaned lambs. Subsequent investigations demonstrated dramatic responses in the live-weight gains of lambs treated with cobalt bullets or dosed with cobalt chloride, although the latter treatment had to be repeated at intervals of about three weeks. The work also showed the usefulness of determinations of a urinary metabolite — formiminoglutamic acid (FIGLU) — in the diagnosis of subclinical cobalt deficiency (Russel, Whitelaw, Moberly and Fawcett, 1975). Later work, again with weaned lambs, showed that serum vitamin B_{12} concentrations reflected the treatment of the pasture more than one year previously with either 2.25 kg or 4.50 kg cobalt sulphate/ha (Whitelaw and Russel, 1979a).

Although cobalt deficiency is known to occur in certain areas of unimproved hill vegetation, most of which have been identified, it is likely that the condition can be induced by procedures such as liming, which frequently precede pasture improvement. Another important deficiency which falls into the same category is that of copper, where the raising of soil pH is known to increase the concentrations of molybdenum and

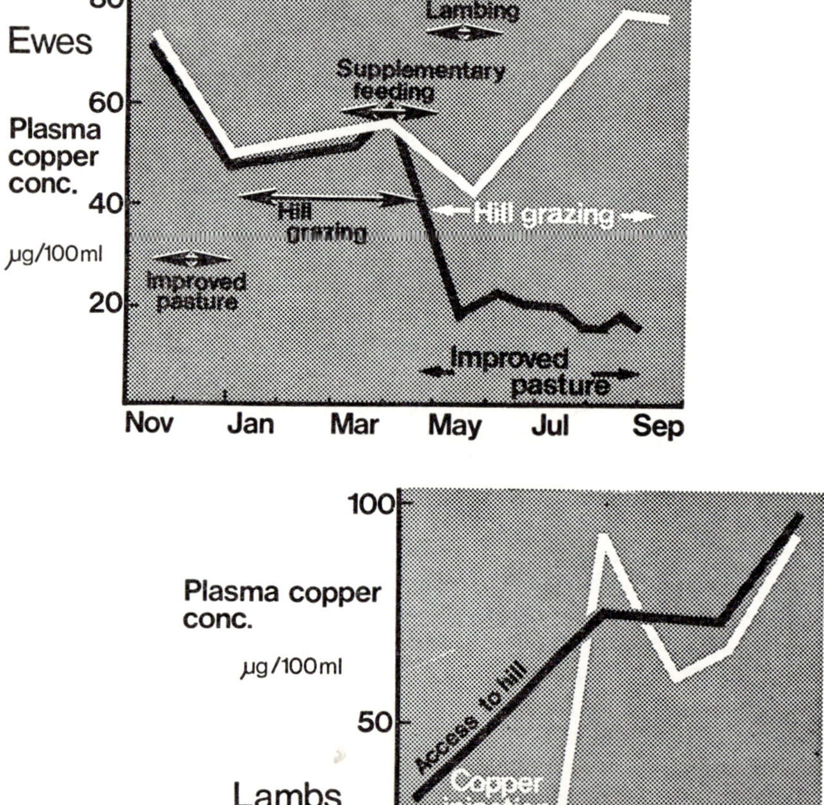

Fig. 8   Plasma copper concentrations in ewes kept wholly on hill grazings or on improved pasture during mating and lactation, and suckling lambs kept wholly on hill grazings or on improved pasture with and without copper therapy.

sulphur in the herbage and, through the interaction of these elements with copper, induce a deficiency of this trace element in the grazing animal.

Recent work at Sourhope has shown that ewes and lambs grazing reseeded pastures during lactation had very low plasma copper concentrations while those of similar animals on the unimproved hill grazing were within the normal range (Fig. 8). An experiment in which one member of pairs of twin lambs was kept in copper sufficiency (Fig. 8) by the periodic injection of copper demonstrated an advantage in live weight at weaning of 2.5 kg as a result of the copper therapy. Differences in various haematological parameters, plasma and liver copper concentrations, fleece quality and skeletal characteristics were also demonstrated (Whitelaw, Armstrong, Evans and Fawcett, 1979). Further studies have shown that copper deficiency in lambs cannot be relieved by the administration of copper to ewes during mid-pregnancy but, given at lambing time, it will increase the copper content of the milk and afford protection to the lamb for 10-12 weeks. After that period further copper therapy is required to maintain satisfactory growth rates and prevent the development of copper deficiency.

Current studies on the problem of copper deficiency are conducted by a group of interdisciplinary research workers.

Animal health

An extensive range of animal health problems associated with hill sheep production has been investigated in the Organisation over the past 25 years. Mention of many of these, such as the work on trace element deficiencies, studies on the premature wear and loss of incisor teeth, work on calcium and phosphorus supplementation and mineral metabolism, factors predisposing hypomagnesaemia and the effect of the administration of selenium, has already been made; others are dealt with later in relation to the veterinary monitoring of the hill sheep development projects (Chapter 6). It only remains here to indicate some of the aspects not covered elsewhere and to acknowledge the Organisation's indebtedness to many of its sister institutes and other bodies for their assistance in dealing with a wide variety of veterinary matters, particularly in the earlier days before the Organisation had its own veterinary section, and had little to offer its collaborators except a long list of apparently intransigent animal health problems.

In the fifties work on helminths in sheep and liver fluke in cattle was carried out at Lephinmore by the West of Scotland Agricultural College, and on liver fluke in sheep at Glensaugh by the Ministry of Agriculture's Veterinary Laboratory at Weybridge. The need for control of both types of endoparasites became more urgent with the intensification brought about by the advent of the systems development projects. Advances in the field of anthelmintics enabled helminths to be effectively controlled by strategic dosing programmes developed at an early stage by the Organisation's veterinary section, but it was not until the seventies, when a flukicide effective against the immature stages of the fluke became available, that liver fluke was virtually eliminated from flocks where previously almost every ewe was affected (Whitelaw and Fawcett, 1977).

The problems of tick-borne fever and tick pyaemia were extensively investigated during the sixties by Dr W M N Foster, who conducted a comprehensive survey of the areas of Scotland where the sheep tick (*Ixodes ricinus*) is found. Dr Foster's work on the relationships between tick-borne fever and tick pyaemia, on the identification of different strains of tick-borne fever, and on the aetiology and epidemiology of tick pyaemia, made an invaluable contribution to knowledge in these fields.

Another major topic is the identification of causes of mortality. In some respects this aspect, although ever present and by no means unimportant, is not now the major factor that it once was. Intensification in any form tends to bring with it the risk of increased losses. However, improvements in management, particularly in nutrition during pregnancy and lactation, which have made intensification possible, have led to significant reductions in the level of mortality in both ewes and lambs. This does not mean that attention to veterinary matters can be relaxed, but it does provide grounds for the expectation of a continued improvement in animal health as ways are found of making further improvements in the management and nutrition of hill sheep.

Condition scoring

It is perhaps pertinent at this juncture to mention one particular technique which has proved extremely valuable and which has been widely used throughout the research programme on the components of production — the simple technique of condition scoring.

Subjective assessments of fatness had been made routinely since recording was initiated on the three research stations, but these had necessarily been very personal and highly variable measurements based on a multiplicity of scales which, in some instances, varied throughout the year with the notion of what was a proper condition for a ewe at that particular time. During Dr Reid's time with the Organisation he introduced a system of condition scoring which had been used in Australia. This was employed initially in some of the pregnancy nutrition studies and later in the reproductive investigations of the effect of nutrition and body condition on ovulation rate. The technique was particularly useful in experiments requiring the production of groups of sheep of uniform body condition, but it was perhaps the demonstration of the close relationship between subjectively assessed body condition and chemically determined measurements of body fat (Russel, Doney and Gunn, 1969) which showed the potential value of condition scoring as a technique in both management and research.

The value of the technique, and the reason for the widespread use which it now enjoys, lies in the clear description of certain physical characteristics identifiable in sheep of different degrees of fatness. This enables it to be learned readily, affords a better description of fatness than terms such as 'fairly good' or 'forward store', and allows an immediate understanding of the quantified descriptions by a wide variety of users. It must, however, be remembered that it is a subjective technique which attempts to categorise a continuous variable in a system of discrete classification, and as such can never be perfect. Its value has nonetheless been amply demonstrated by its adoption and widespread use throughout the country by the Meat and Livestock Commission and other advisory and development services.

The present position

The research work carried out in the Organisation during the last 25 years on the many aspects of hill sheep production has confirmed and underlined the importance of nutrition which was recognised in a general sense, but not fully understood, at the outset. The studies conducted on the various components of production, the integration of the results of these separate research topics into our present view of year-round nutrition, complemented by the insights provided by the programmes of research on specific topics such as supplementation, the regulation of food intake, and the role of certain minerals and trace elements, combined to give a reasonable understanding of the quantitative relationships between nutrition and production, and how these may be manipulated to improve the efficiency of production.

It would, however, be misleading to give the impression that progress towards an improved efficiency of production can be made only through research on the many facets of nutrition. As already indicated, for example in relation to the discussion of the persistency of lactation, the choice of genotype is also important. The nutritional research work has shown clearly that the potential production of hill sheep, and in particular the reproductive and lactation potential of the Blackface ewe, is considerably greater than would at one time have been supposed. Even in the most advanced of the development projects individual animal performance does not begin to approach this potential. Indeed, the complex balancing of output per ewe and number of ewes, which is more properly considered in relation to the hill sheep development studies (Chapter 6), dictates that this is so. This does not mean that existing breeds must be retained at all costs until individual output attains or closely approaches the potential production of that genotype (Russel, 1978a). In recent years the performance of other genotypes, notably Border Leicester, Texel and East Friesland crosses with the Blackface, has been examined in anticipation of the need to effect a breed change at some future date. The work on upland sheep (Chapter 7) has also provided much valuable information in this respect.

It would also be misleading to give the impression that the work on nutrition and production of hill sheep over the years has resolved all the problems and answered all the questions in these areas. Much is known and understood, but much also remains to be done. That the identification of the areas requiring further research, and the ordering of research priorities pose few problems is due in no small measure to the complementary nature of the Organisation's research and development activities. While these functions proceed in parallel, it would be wrong to imply that the two never meet; in practice their respective paths cross and interlink at many points. Research is commonly thought of as leading on to development, but it is our experience that the converse is equally true, and there are numerous examples of current research topics which have been identified as the result of an examination of on-going development projects; the present work on mid-pregnancy nutrition is one such case.

In some measure the nature of much of the nutritional research has changed in recent years. There was a period when it appeared that most research topics proceeded through three clearly identifiable stages — a period of making measurements in the field to characterise the problem and to obtain basic quantitative information, such as food intakes or live-weight changes, followed by a period of experimentation under the closely controlled conditions of the animal house, metabolism room and laboratory, followed in turn by a final phase in which the research results were taken back to the field for testing. This is no longer true, partly because many situations are now well characterised and much of the first-stage quantitative information now exists, but also because the truly experimental or research stage can no longer always be conducted in the animal house or metabolism room. The work on mid-pregnancy nutrition and its interaction with nutrition at other times of the year, and the programme on the supplementation of grazed herbage, are examples of projects which require that the basic experimentation — and not just the final testing of research findings — be conducted in the field. This requirement brings with it a new set of technical problems concerned with the means of making the relevant measurements in what are very often difficult and demanding situations. It will be interesting to read in another 25 years' time of the outcome of our present ideas on how these problems may best be tackled, to learn of the solutions to what appear to be major problems now, or indeed even to learn whether the same problems remain or if they have been replaced by new ones.

5

Grazing ecology

Contributors: Rhd H Armstrong, J Eadie, M J S Floate, S A Grant, R F
Hunter, T J Maxwell, J A Milne, A R Sibbald, D A Vine

Co-ordinator: J Hodgson

Introduction

In the First Report of HFRO (1954-58) the aim of the then Botanical
Studies Department was stated to be 'to devote attention to finding ways
of improving the utilisation and management of the hill vegetation itself'.
It was considered that the programme should include studies of the ecology
and response to grazing management of each of the main hill plant communities, as well as studies of the relationships between grazing animals
and their diverse grazings.

Studies centred on these broad objectives form the main theme of this
chapter although, with time, increasing attention has been devoted to
parallel work on sown swards. This reflects the progressive development
of interest in a research programme set in the context of the whole farm,

and therefore concerned with the optimum use of all the alternative vegetation resources available.

The work described falls under two main heads:

(a) The influence of the grazing animal on the botanical composition of the sward and on herbage production, including the effects of selective grazing and of nutrient cycling via urine and faeces.

(b) The influence of sward characteristics on diet selection and nutrient intake by the grazing animal.

In each case there has been a major interest in the degree to which the interactions between sward and animal can be modified to advantage by control of grazing. The material has been arranged as far as possible to illustrate the relationships between the two main areas of interest, and to indicate the way in which the evolution of thinking on grazing management has influenced the development of the research programme.

First phase: observation and deduction

In the early days attention was concentrated primarily on descriptive studies within the context of traditional set-stocked, year-round grazing systems. Studies of the *Festuca-Agrostis* grassland complex in south east Scotland (King, 1962) and of upland heath in north east Scotland (Nicholson and Robertson, 1958) allowed some deductions to be made on the role of biotic factors in the formation of the existing vegetation. Further phytosociological studies were made of hill grasslands throughout the south of Scotland, the Ochil Hills and the Highlands south of the Great Glen. This work, together with that of many other ecologists, was reviewed comprehensively by King and Nicholson (1964), who gave a detailed analysis of the nature of variation in the vegetation in relation to variation in soil moisture regimes, soil base status and grazing history. The results of all these studies gave a strong indication of the influence of selective grazing and grazing pressure on the floristic composition of hill swards.

King and Nicholson (1964) considered that 'in the planning of land use programmes and, in particular, for land of inherent low fertility, further work is necessary on the conception of soil-plant-animal systems not only from the standpoint of their impact on internal processes but also from the point of view of their relative efficiencies in controlling the nutrient input:output ratio in relation to all sources of gain and loss. Furthermore, it is often too readily assumed that the use of the forest zone for animal production is unsound land usage, but many of the factors on which such a conclusion is based may be the result of current management practices rather than the effects of sheep grazing *per se*. The lack of grazing control inherent in present practice and the need for frequent burning results in vegetation trends of an undesirable kind with reactions throughout the entire system'.

Within this statement are implicit a number of suggestions which have a bearing on the development of the research programme at HFRO. First, reference is made to the view common at the time that sheep grazing had led to the deterioration of hill lands; second, attention is drawn to the lack of grazing control in contemporary management practices; third, further work is called for on nutrient balances in soil-plant-animal systems; fourth, it is implied that, because changes in one part of the system have consequences throughout, the ecosystem approach to research on the whole system should be adopted.

Ecological studies of traditional grazing systems

Hunter (1960a) recognised that improved economic exploitation of the hills could only be rationally founded on an adequate understanding of the ecology of traditional management systems. Throughout this early period, he and his co-workers were engaged in detailed observations on the ranging habits of sheep on an enclosed hillside on the Sourhope Research Station (Hunter, 1962a; Hunter and Davies, 1963; Hunter and Milner, 1963). These studies provided important information on the grazing habits of sheep and the seasonal patterns of use of a range of plant communities (Hunter, 1962a) which formed the basis for much subsequent grazing research. Information derived from these same studies demonstrated clearly the relatively limited movement of family groups of sheep, and indicated the grounds for a relationship between the area on the hill where the sheep was born and its subsequent productive performance (Hunter, 1964a).

There are marked contrasts in soil and vegetation within relatively small areas of hill ground. Hunter's (1962a) major study clearly showed that sheep distinguish between those swards which occur on brown earth, basic gley, and podsolised brown earth soils (mull) and those growing on humus podsols, podsolic acid gleys and peat podsols (mor) (Chapter 2). The former, primarily communities of *Agrostis* and *Festuca*, are grazed comparatively heavily, and the latter, characterised by communities dominated by *Nardus, Molinia, Calluna* or *Eriophorum*, are grazed relatively lightly. Hunter pointed out that in these conditions grazing is not a primary determinant of vegetation change, but merely hastens and accentuates the influence of the climatically determined soil type on the vegetation. However, the long-term effect of uncontrolled grazing is to reduce the grazing value of a hill pasture. He argued that sustained improvement in production could only be achieved by breaking the free relationship between sheep and vegetation, and introducing some form of grazing control. Furthermore, if grazing control was to be used for improvement, the primary division had to be between *Agrostis-Festuca* swards and grass heath or shrub heath communities, otherwise the pasture-animal relationship associated with free grazing would simply be maintained, albeit on a smaller scale.

The information from these detailed studies provided strong indications of differences in nutrient intake between seasons of the year, and between individual sheep grazing the same hill. It had long been recognised that hill pastures provided a relatively poor nutritional environment for ruminant animals, but an examination of the implications of Hunter's evidence only became possible when the results of contemporary studies on the influence of nutrition on the components of sheep production became available. This work is reviewed in Chapter 4. For present purposes it suffices to say that improved performance required a general improvement in nutrition from pasture, with particular importance being attached to the period of lactation and to autumn and early winter nutrition.

An obvious limitation of the sheep behaviour studies lies in the fact that they provided no information on levels of herbage intake by grazing sheep. The need for such information was becoming increasingly clear. On the one hand the grazing behaviour work had given a broad indication of possible improvements in hill pasture utilisation, although both the analyses of existing traditional systems and the more precise formulation of ideas about improvement required information of a directly nutritional kind. On the other hand, the developing preoccupation with the components of sheep performance began to highlight the need for more precise and direct information on the levels of nutrient intake obtained by hill sheep from their pasture.

It was against this background that a study was made in the years 1961-64 of the annual cycle of nutrient intake by sheep, using Hunter's original study area at Sourhope (Eadie, 1967b). The procedure employed was to run with the ewe stock a small group of wethers which were born on the unit, and to make faecal collections from these animals on three consecutive days each month throughout the year. The digestibility of the ingested herbage was predicted from faecal nitrogen concentrations, using regression equations derived from over 50 digestibility studies on herbages from the range of pasture types occurring on the unit. The annual cycle of ingested herbage digestibility is shown in Fig. 9.

As a result of this work it was possible to make an analysis of the consequences of the traditional management system in the following terms (Eadie, 1967b):

Hill sheep are normally set-stocked in year-round, free-range grazing systems. Hill pasture growth on the other hand is highly seasonal. One consequence is that over the period of active pasture growth the amount of herbage eaten is small in relation to what is produced. Thus the sheep

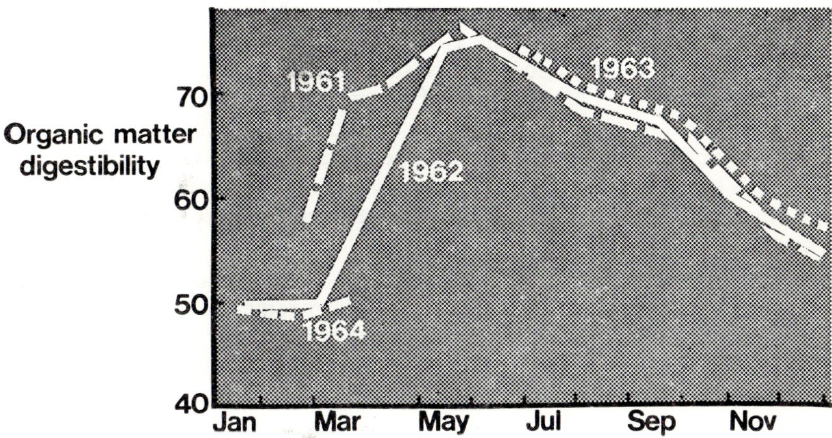

Fig. 9   Seasonal changes in the digestibility of the diet consumed by sheep set-stocked on a Cheviot hill (1961-64).

are able to graze selectively and maintain the quality of their diet. But the uneaten herbage, which is a large proportion of the total production, accumulates *in situ,* matures, and so dilutes the quality of the available feed. Deterioration continues throughout the winter as a result of selective grazing activity and the death and decay of plant materials, but there is a considerable carry over of dead herbage from one season to the next.

Traditional hill management systems therefore create a vicious circle. Stocking rates, set at levels determined by the need to maintain a certain minimum level of winter nutrition, allow the sheep to graze selectively. This process appears to buffer the animal against a rapid fall in the quality of the diet during the summer, but also creates a fund of mature herbage which limits diet quality in early summer.

These arguments about the ways in which traditional grazing management might limit animal performance had to take into account Hunter's conclusion that pasture improvement by grazing control required the separation of the *Agrostis-Festuca* and heath communities. It was therefore important to learn something about the nutritive value of the species and plant communities characteristic of these vegetation types, and their responses to grazing manipulation. The development of the *in vitro* digestibility procedure at this time (Tilley and Terry, 1963) made it possible to study individual hill pasture species. Black (1967) demonstrated that a number of these species could provide material of high digestibility in spring, and of quite reasonable quality throughout the summer, if managed in an appropriate fashion.

Parallel studies of the nutritive value and utilisation of an *Agrostis-Festuca* sward demonstrated the difficulty of combining high levels of utilisation and good quality ingested herbage in the presence of the amount of dead herbage (approximately 2000 kg DM/ha) normally found on these swards. However, by reducing this fund of dead herbage to about 600 kg DM/ha the efficiency of utilisation of organic matter at subsequent grazings could be doubled, whilst at the same time the digestibility of the herbage consumed could be increased by between 4 and 10 units (Eadie and Black, 1968). This series of studies also gave evidence of a considerable potential for improved autumn nutrition by the controlled grazing of *Agrostis-Festuca* swards.

The *Nardus-* and *Molinia*-dominant pasture types, on the other hand, were shown to be much less amenable to improvement by grazing control. The fund of under-utilised herbage was much greater; for example, over 3500 kg/ha in the *Nardus* sward studied by Eadie and Black (1968). Such swards only produced about 1200-2000 kg/ha/yr compared to about 2200-3200 kg/ha/yr for *Agrostis-Festuca* swards. In addition, the herbage from traditionally managed *Nardus* and *Molinia* swards was shown to be several percentage units lower in digestibility than that from *Agrostis-Festuca* swards. It was also apparent that herbage even from well grazed *Nardus* and *Molinia* swards would be less digestible than that from *Agrostis-Festuca* swards because the digestibility of the green material was lower (Eadie and Black, 1968).

Thus the arguments all pointed to the need for a degree of grazing control in hill management. In the hill environment in south east Scotland, grazing control could be employed to improve the utilisation and nutritive value of *Agrostis-Festuca* swards, which often amount to between 15% and 40% of a hill grazing unit. The better quality herbage could be used to provide better nutrition during the lactation and lamb growth period, and, following a mid-season rest, in the pre-mating and mating periods. Better overall pasture utilisation would require increases in stocking rate, and the consequences of such increases, together with any improvement in ewe fertility, would require the provision of late pregnancy supplements on a regular basis.

During this period it was realised that the synthesis proposed for hill environments which included a useful proportion of *Agrostis-Festuca* sward could be generalised. Although other hill environments did not have indigenous vegetation capable of responding to grazing control in the same way as *Agrostis-Festuca,* there was a wide range of hill land improvement techniques which could fulfil, albeit at much greater initial cost, the role envisaged for managed *Agrostis-Festuca* in the proposed system. Thus it was possible to circumvent the limitations to overall stocking rate which, as Hunter saw, were set by the proportion of mull soils on a hill. There was nothing new in the idea of hill land improvement, but here was a clear strategy, based on firm evidence, which promised to make the most effective use of expenditures on hill land improvement (Eadie, 1970a).

The subsequent development of a system of hill management based on these concepts is described in Chapter 6, and the results of work on alternative land improvement techniques are outlined in Chapter 3.

The grazing influence on hill vegetation

The arguments advanced by Hunter (1962) and Eadie (1967b) justified grazing control in terms of improvements in nutrient intake by grazing animals. The consequences of intensified grazing on the vegetation and on levels of plant nutrients in the soil were unknown, but studies of vegetation contrasts on either side of long established fence-lines in many parts of Scotland had demonstrated the nature and magnitude of the divergent developments which can occur from common origins (Nicholson, 1967). These contrasts were often between *Calluna*-dominated communities on the one hand, and some form of grass heath on the other, where the differentiation in site characteristics resulted from apparently slight differences in use over long periods. Associated soil differences between these paired sites were studied (Chapter 2) and it was concluded, contrary to popular beliefs in the degenerative effects of sheep grazing, that a slightly increased animal influence — possibly through long-term enhanced recycling of plant nutrients — had brought about some improvements not only in vegetation but also in soils. These effects had, however, been observed under conditions of relatively uncontrolled grazing, and the results of greatly intensified animal influence under controlled grazing had not then been investigated.

At this time studies were also made of the variation in yield, date of flowering, winter greenness and sensitivity to cutting of hill grass species (Grant and Hunter, 1968b). This work indicated that the response of natural hill pastures to seasonal variations in grazing intensity would be greatly affected by their botanical composition. Investigations of the dynamics of *Nardus* pasture (Nicholson, Paterson and Currie, 1970) included both clipping and grazing experiments. Clipping experiments showed the inability of *Nardus* to withstand repeated close defoliation. However, in experimental plots it was difficult to achieve close grazing of *Nardus* owing to the marked preference shown for other plants. Cattle grazed *Nardus* earlier in the year than sheep, but also grazed the plants less closely than sheep. Floate, Eadie, Black and Nicholson (1973), in an extension of this study, concluded that the improvement of *Nardus* pasture by grazing control alone would incur unacceptable penalties to animal performance.

Muirburn

In contrast to the evidence of the intensive grazing of *Agrostis-Festuca* grassland by sheep (King, 1962; Hunter, 1962) on *Molinia*-dominant pasture the deciduous nature of the season's growth, coupled with low levels of utilisation and decomposition, leads to a build up of uneaten litter which impedes access to fresh growth. In heather *(Calluna vulgaris)* dominant areas the woody framework of the plant develops more quickly when the plant is undergrazed and the stand rapidly reaches heights and densities which impede sheep movement. Periodic burning removes the accumulated material and restores the grazings to a condition where the fresh young shoots are more accessible to the grazing animals, but considerable controversy has existed amongst ecologists, farmers, gamekeepers and managers about the benefits and dangers of this practice. It was suggested, for example, that continued muirburn and sheep grazing extracted nutrients and led to a general degrading of soil fertility, but little factual information was available on losses of plant nutrients, the size of the nutrient fund in soil and vegetation, and on the factors influencing the regeneration of vegetation after burning.

The common interest and concern of scientists from agricultural, conservation and academic backgrounds led in the early sixties to the formation of a Muirburn Committee, all of whose members were actively engaged in research on the topic. This Committee met at intervals of two to three years during most of the sixties to discuss problems and report progress. The HFRO contribution to this concerted effort consisted of a joint study with the Macaulay Institute for Soil Research on the nutrient fund in heather ecosystems (Robertson and Davies, 1965), a survey of heather regeneration at 30 sites in Scotland which were burned as part of the normal farm/estate management (Grant, 1968a), an experiment to investigate the interaction between grazing and burning on heather moors (Grant and Hunter, 1968a; Grant, 1971b), and a study of the effects of muirburn on *Molinia*-dominant communities (Grant, Hunter and Cross, 1963). To aid the interpretation of the heather grazing and burning experiment, investigations were made of the effects of frequency and season of clipping on the morphology, productivity and chemical composition of heather (Grant and Hunter, 1966). This followed a study of the ecotypic variation in *Calluna* (Grant and Hunter, 1962).

These studies showed that grazing reduced the weight of vegetation on heather stands by suppressing the build-up of woody stems, so that grazed heather was short and dense in comparison with ungrazed heather. However, levels of utilisation approaching 60% of the current season's shoots did not affect growth in the following year. Sheep grazed stands of this type, or those recently burned, in preference to older heather. New shoots on grazed heather were richer in nitrogen and minerals than shoots on adjacent ungrazed heather of the same age.

Later, following submissions made to the Select Committee on Scottish Affairs on Land Resource Use in Scotland (session 1971-72, Sub-Committee A — Rural Land Use), it was recommended that the regulations on muirburn be examined and that means should be sought to promote the best possible practice. Discussions took place between the Department of Agriculture and Fisheries for Scotland (DAFS) and the Nature Conservancy Council (NCC) and culminated in the setting up of a working party, including HFRO representation, to evaluate existing knowledge and note any needs for further research, and to produce a guide to good muirburn practice.

This guide was published in 1977 as an HMSO booklet (DAFS/NCC, 1977) and while it was intended primarily for use by those practising muirburn it also provides a concise theoretical and practical basis for students or advisers wishing to study this management practice in more detail.

Second phase: consolidation

The strategy of grazing control postulated by Eadie (1970a, b, 1971) and outlined earlier in this chapter provided a means of improving the nutritive value of herbage on selected parts of the hill by reducing the dead component, and of improving nutrient intake by the grazing sheep at critical times of the year. This resulted in a marked change from the original approach, in which pasture studies were carried out against the background of the assumption that grazing sheep had free access to all parts of the hill throughout the year, and influenced the course of a number of aspects of the research programme in the late sixties and early seventies.

First, the concepts of pasture improvement, coupled with the results of studies on the influence of nutrition on ewe lactation and lamb growth (Chapter 4), highlighted the need to examine the potential for improving ewe and lamb performance on improved pasture. Second, although the effects of this strategy on the sward were largely unknown, it was recognised that many of the undesirable features of acid grasslands had arisen as a result of underutilisation, and it was therefore considered that increased grazing pressures were unlikely to do them much harm. The main research interest for these pasture types centred, therefore, on the effects of increased grazing pressure on botanical composition and nutrient cycling, and hence on herbage production and nutritive value. There was also an active interest in alternative methods of improving enclosed pastures. In the more vulnerable vegetation types such as blanket bog and heather moor, however, priority was given to studies on the possible adverse effects on botanical composition and herbage production of increases in grazing pressure which, in the context of the postulated system of grazing control, would be experienced in summer and winter.

These aspects of the programme are considered in turn.

Lactation and lamb growth at pasture

Several experiments were carried out at Sourhope with lactating ewes and their lambs, or weaned lambs, in which observations on herbage consumption were combined with measurements of animal performance. Parallel indoor experiments with animals fed on cut herbage provided estimates of herbage digestibility, and acted as checks on the field measurements.

In each of two successive years, ewes grazing an old reseed dominated by *Agrostis* spp. and *Dactylis glomerata,* which had been closely grazed the preceding autumn to remove dead herbage, ate 17% more herbage dry matter, with a digestibility four percentage units higher, than ewes grazing a traditionally managed *Agrostis-Festuca* sward containing approximately 2500kg DM/ha of dead material. This resulted in improvements in milk yield of 20% and 75% for ewes nursing singles and twins respectively. The magnitude of this response clearly illustrated the importance of relatively modest increases in the digestibility of ingested herbage.

At this time there was an active interest in the potential contribution of white clover to the enhancement of nutrient intake, since in the indoor trials this effect was shown to be greater on *Agrostis-Festuca* than on ryegrass herbage. Other work was concerned with the value of milk in the diet of the grazing lamb, since preliminary studies indoors and at pasture showed that milk appeared to act as a true supplement, and that the effect was largely independent of herbage quality (Armstrong and Eadie, 1973a, b). However, these results were not followed up at the time, partly because of the need to refine techniques to measure the intake of mixed diets under grazing conditions, and partly because of the movement of the staff concerned from Sourhope to headquarters in Edinburgh.

Grazing management and nutrient cycling

One of the major limitations to herbage production is the availability and continuity of supply of plant nutrients in soil. Where little or no fertiliser is applied, the rate of organic matter decomposition and the release of nutrients by mineralisation are controlling determinants of supply (Chapter 2). In traditional systems the plant litter return pathway is likely to be by far the most important for nutrient return, whereas intensified grazing and increased utilisation leads to a greater return of nutrients via the animal pathway (Floate, 1970d, e).

In the sixties studies were undertaken to compare the rate of release of nutrients from materials of plant and animal origin (Floate, 1970a, b, c; Floate and Torrance, 1970). Herbages cut from *Agrostis-Festuca* and from *Nardus*-dominant pastures, either cut frequently to simulate frequent defoliation by grazing animals or allowed to accumulate over the growing season before cutting in the autumn, were subsequently fed to housed sheep in digestibility trials. Total herbage dry matter production was lower on the plots cut frequently than on those on which herbage was allowed to accumulate, but the total uptake of nitrogen and phosphorus was greater because of their higher concentrations in the fresh green material under frequent cutting (Floate, 1970d). When samples of the herbages and of the faeces derived from them were incubated under temperatures and moisture regimes similar to those on hill pastures, mineralisation of nitrogen and phosphorus was greater from the faeces derived from frequently cut herbage than from the annually accumulated herbage itself, and in fact immobilisation of these nutrients often occurred in the plant tissue.

The major difference between the plant litter and animal excreta return pathways lay in the digestive processes which occur within the grazing animal. These processes lead to the release of large amounts of readily available nitrogen in urine, in amounts proportional to the digestibility of the ingested herbage, and to the concentration of phosphorus in the faeces with a consequent increase in the organic (potentially available) component of the total. There was an approximately ten-fold increase in potentially available nitrogen and phosphorus via the animal return pathway on both pasture types (Floate, 1970d, e, 1971) (Fig. 10). Laboratory evidence was thus obtained for the potential benefits to herbage production of increased intensity of grazing. However, the term 'potentially available' is used advisedly because under grazing conditions losses may be sustained in a number of ways including leaching, volatile loss, and uneven distribution of excreta on pasture, and because nutrients may be inaccessible to plant roots.

In experiments conducted by Black and Floate, and described by Floate (1970e), measurements were made of the production responses of *Agrostis-Festuca* pasture to application of sheep faeces and urine. It was shown that hill pasture responded to the increase in the amounts of available nitrogen and phosphorus in excreta, and that the response to urine nitrogen (up to 20 kg DM/kg of nitrogen applied) was not less than the response to fertiliser nitrogen. After treatments had been continued for three years, soil analysis showed a small increase in soil nitrogen under full return treatments and a small decline under treatments simulating 50% utilisation of herbage with only faecal returns. These results have recently

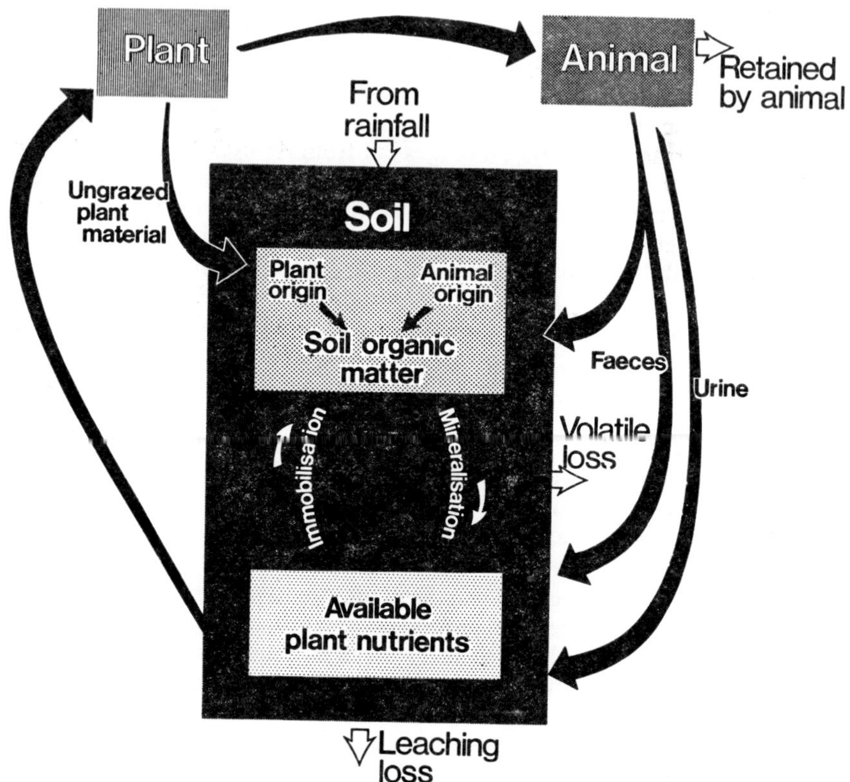

Fig. 10   Simplified nitrogen cycle in soils under ungrazed and grazed swards.

been interpreted to provide evidence that nitrogen may be recycled through the soil-plant-animal system more than once per season (Floate, unpublished). Such a process would lead to more efficient production from limited supplies of available nitrogen in hill soils.

In grazing experiments set up to study pasture dynamics (Nicholson *et al*, 1970) it was shown that, compared with the low level of nutrient cycling under extensive hill management, surface improvements and increased herbage utilisation resulted in an increase in the cycling pool of nitrogen from 0.5 to 5.0%, and of phosphorous from 0.4 to 2.5% of the total soil pool of these nutrients, together with enhanced plant and animal production (Floate *et al*, 1973). However, the treatments in these experiments were originally designed to study pasture dynamics, and were not necessarily appropriate to the developing philosophy of the requirements of improved pasture in the systems development programme (Chapter 6).

Nicholson (1968) and Nicholson, Currie, Paterson and McCreath (1968) advocated a policy of grazing control combined with a limited patchwork of land improvement for areas of low quality native vegetation, in the expectation that grazing would improve the pasture and enhance its capacity for future production as vegetational succession took place round the perimeter of the improved areas. It was envisaged that these trends would be consolidated by the radial spread of plant nutrients in urine and faeces from these nuclei. A test of this hypothesis was set up at Lephinmore but subsequent changes in the management of the experimental areas precluded examination of the effects of the patchwork improvements. It is of interest, however, that the swards sown in 1958 and 1959 have been maintained up to the present time with very limited inputs of artificial fertiliser.

Grazing control and pasture improvement

It was anticipated that, arising from the development work in systems studies, information would be required on the best choice of soil and pasture type for improvement, and on the choice of appropriate combinations of improvement techniques. Accordingly, starting in 1969, a programme of evaluation of a range of improvement treatments on different soil and pasture types was undertaken at Sourhope. These large plots, grazed according to a defined management strategy appropriate to their use in a two-pasture system (Chapter 6) involved three different vegetation types on separate sites, each with five treatments ranging from grazing control alone to full surface reseeding with grazing control. A fundamental feature of these improvement treatments was that grazing control would be exercised in an attempt to improve pasture composition, to reduce the dead:green ratio of available herbage, and to enhance nutrient recycling via grazing sheep.

These experiments provided an opportunity to study the long-term effects of nutrient recycling under a range of treatments on contrasting soil and vegetation types from *Agrostis-Festuca* on brown forest soil, to *Nardus*- and *Molinia*-dominant grass heath on peaty podsol. Preliminary results show that, progressively with time within each treatment, and progressively across the range of increasingly comprehensive treatments, there was a significant increase in the amounts of major plant nutrients recycled in the form of excreta from grazing sheep. These increases were not only related to increases in herbage production, but also represented the utilisation of an increasing proportion of the total soil nutrient pool.

These experiments have shown that grazing control, increased herbage utilisation by grazing sheep, and enhanced nutrient recycling, together with nitrogen fixation by clover on some treatments, result in the more efficient utilisation of soil nutrient supplies for animal production. The effects are progressive, and the experiments are continuing in order to measure long-term effects on herbage and animal production.

Grazing studies on blanket bog

A study was set up in 1971 to provide background information on the consequences of changes in management associated with the introduction of two-pasture systems on blanket bog. Three stocking rates were provided at each of three sites at Lephinmore. Seasonal patterns of utilisation of the individual species were monitored, as were changes in the floristic composition, morphology and seasonal patterns of herbage growth in the swards. Samples of bog species were also collected for estimation of nutritive value. The results (Grant, Lamb, Kerr and Bolton, 1976; Grant and Campbell, 1978) were discussed in relation to systems of sheep farming and the nature of vegetation mosaics in hill land, and it was argued that the lack of potential for regrowth during late summer and autumn would have the result that the quality of grazing afforded by bog vegetation in autumn and winter would be poor whatever the management employed.

The assessment of the effects of the different levels of grazing on production from these bog communities is still in progress; this is necessarily a long-term study if stocking rates are low enough to be realistic. The rates employed, the equivalent of 1.5, 1.0 and 0.5 ha/sheep, result in the utilisation of 10-15% of the annual dry matter production at the lowest stocking rate, and two and three times this amount at the intermediate and high stocking rates respectively. After six years of grazing there were no differences in current season's dry matter yields (up to the end of July/beginning of August) between plots at the lowest and intermediate stocking rates. However, at the highest stocking rate yields were reduced by some 25%. Effects on individual species are less clear because of seasonal fluctuations in floristic composition and initial differences in composition amongst the three sites.

Grazing studies on heather moor

The work on heather moor began a new phase in grazing research within the Organisation, involving close collaboration between scientists with interests in sward responses and animal nutrition respectively, and therefore provided a forerunner to the interdisciplinary grazing studies which followed on a range of vegetation types.

As long ago as 1953 it had been concluded in the Second Report of the Scottish Hill Farm Research Committee that there was a need for more precise information on heather as a food for hill sheep. It was recognised then that a knowledge of the chemical composition of the heather plant was impossible to apply to a consideration of problems in animal nutrition unless the samples analysed represented what a grazing sheep would normally select.

It was 17 years before this recommendation was acted upon. In that time the only information on diet selection or the nutritive value of heather came from a descriptive botanical analysis of the rumen contents of sheep at Lephinmore by Macleod (1955) and a feeding trial with dried material by Armstrong and Thomas (1953) at King's College, Newcastle. Current season's shoots of heather from the long-term experiment of Grant and Hunter (1966) were analysed, using the *in vitro* digestibility method, but it was concluded by Grant and King (1970) that the estimates of digestibility obtained were not indicative of the true *in vivo* value of heather.

Renewed interest in the selective grazing and intake of heather by the grazing animal arose as a result of the coincidence of several factors. In the late sixties hill sheep farms on predominantly heather hills appeared particularly vulnerable to economic hardship. The two-pasture system had been developed in relation to predominantly *Agrostis-Festuca* vegetation. However, before the principle could be applied to heather hills it was necessary to know more about the factors influencing the nutritive value and intake of heather. For example, though earlier work by Grant and Hunter (1966, 1968a) suggested that higher levels of utilisation of heather than those found on most hills could be achieved without damaging plant growth, the effects of increasing the level of utilisation on the nutritive value of the diet selected were not known. For these reasons, and because of the advent of new techniques to measure the intake and digestibility of the diet of the grazing animal, a new project was set up to provide more precise information on heather as a food for sheep.

The use of a modified small flail harvester and the immediate cold storage of the harvest of predominantly current season's shoots avoided the problems found in earlier pen-feeding studies with dried material, and provided a useful basis for studies on the nutritive value of heather (Milne, 1974). Sheep fistulated at the oesophagus were then introduced to allow the collection of samples of the material eaten by grazing animals. Determinations of *in vitro* digestibility were made on the samples collected, and used to predict the *in vivo* digestibility of the heather consumed (Milne, 1977a).

These techniques made it possible to set up joint grazing studies in 1972, building on the results of the earlier work and designed to examine the effects of patterns and levels of utilisation in summer and autumn on the growth and morphology of heather, and on the intake and digestibility of the diet selected. The mode of defoliation, amount of overwintering green shoots and carbohydrate reserves were shown to contribute to differences in regrowth resulting from different seasonal patterns and levels of use, and limits to use set by effects on plant production were demonstrated (Grant, Barthram, Lamb and Milne, 1978). The removal of 40% of the current season's shoots did not affect shoot production, but the removal of

80% of current shoots reduced production by half. Heavy autumn grazing was particularly damaging.

Much information was obtained on the factors affecting diet selection within heather communities (Milne, Bagley and Grant, 1979). Up to 40% of current season's shoots could be removed by grazing without reducing the nutritive value of the diet selected, but on its own heather provided a diet which did little more than maintain the live weight of sheep even in summer, mainly because of low heather intakes. Thus, although the levels of utilisation of heather, a relatively cheap source of nutrients, could be increased from 5-10% of current season's shoots up to 40%, the nutritional limitations meant that such levels of utilisation could only be sustained by lactating ewes in summer if heather was a small proportion of the diet. The creation of a fenced mosaic area in which small areas of grass reseeds were associated with heather-dominant areas appeared the most likely means to achieve adequate levels of nutrition for the ewe and lamb in summer and autumn. This reasoning provides an interesting contrast with the arguments used earlier by Nicholson (1968) for a similar solution to the problems of improving the utilisation of inherently poor quality vegetation.

To provide a factual background to the proposals for mosaic reseeds, information was needed on the factors influencing the amount of heather and grass eaten in combination by the grazing animal. To this end a technique was developed, based on the use of a simple phenol, orcinol, which is found in heather but not in grass and is quantitatively excreted in the urine (Martin, Milne and Moberly, 1975; Chambers, White, Russel and Milne, 1976). When sheep were allowed free access to adjacent areas of grass reseed and heather, the proportion of heather in the diet was inversely related to the quantity of herbage available to the sheep (Fig. 11). It was concluded that, with grass occupying 30% of the total area in a mosaic of grass and heather, levels of utilisation of up to 40% of current season's shoots of heather between June and October could be achieved if the weight of herbage on the grass areas was maintained between 500 and 800 kg DM/ha. The resulting diet was estimated to provide adequate nutrition for a ewe with a single lamb in late lactation, and for a modest weight gain in the ewe between weaning and mating.

Thus it is envisaged that a mosaic of grass reseed and heather, fenced off from the rest of the hill and used for ewes in late lactation and after weaning, will improve the utilisation of heather substantially without incurring the nutritional penalties to the grazing sheep which normally ensue. This mosaic approach is now being tested in a full-scale systems experiment at Glensaugh, involving variations in the size and shape of areas of reseed. The patterns of heather utilisation and the nature of vegetational succession in the vicinity of the boundaries between grass and heather are being recorded, in addition to the effects on ewe and lamb performance.

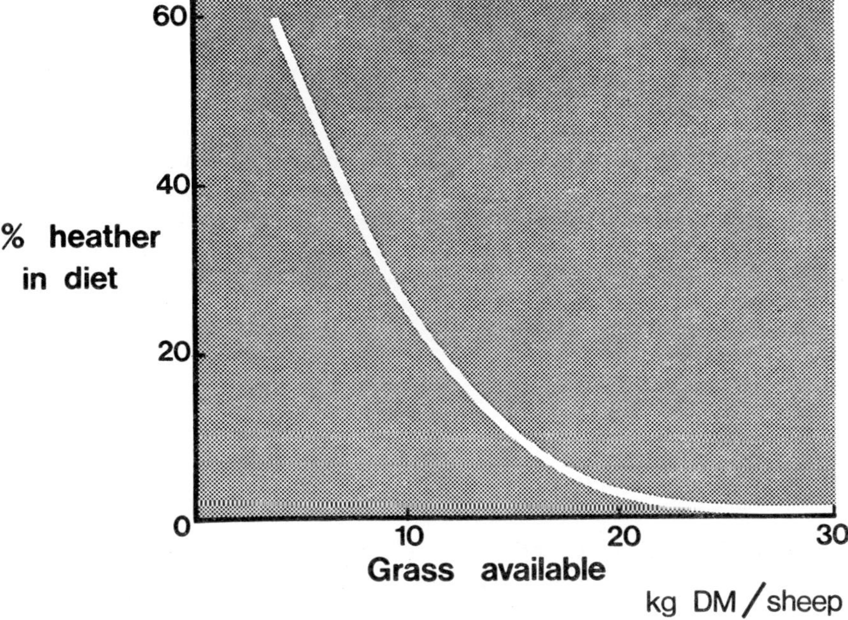

Fig. 11   The relationship between the proportion of heather in the diet (%) and the amount of grass available (kg DM/sheep).

Third phase: interdisciplinary studies

In recent years increasing emphasis has been placed on collaborative grazing studies involving members of the Animal Production and Nutrition Department and the Plants and Soils Department. A major objective of these studies is to increase understanding of the cyclic relationships between the grazed sward and the grazing animal, with a view to improving the efficiency and predictability of systems of management for hill and upland swards. Attention is concentrated at the sward-animal interface, involving studies on (a) diet selection and herbage intake by grazing animals, patterns of defoliation and damage in grazed swards, and the way in which these variables are influenced by sward characteristics (floristic and morphological composition; sward structure and herbage weight) and grazing management, and (b) regrowth of the sward in sufficient detail (for example, changes in sward morphology and structure, carbon exchange, leaf and tiller dynamics) to enable interpretation and prediction of herbage production responses to grazing management. The same basic approach has been adopted in studies on enclosed grass and grass/legume swards

93

where close grazing control is possible, and on indigenous hill plant communities where usually it is not.

Early in 1976, the convergence of interests in aspects of grazing ecology led to the formation of a Grazing Studies Group within the Organisation, comprising members with interests in aspects of diet selection and herbage intake by grazing animals, the effects of grazing on plant growth and morphology, their interactions in the context of grazing systems and the use of models as a means of understanding the processes involved. Collective discussions on the development of a conceptual and, where possible, quantitative framework have enabled each contributor to identify more clearly gaps in knowledge and to gain a greater insight into the relationships between the different disciplines involved in studies of whole grazing systems.

Grazing models

Mathematical modelling was first used in the Organisation in 1973 as a means of investigating the effects of alternative grazing strategies on *Agrostis-Festuca* swards for improved systems of hill sheep production, because of the need to test many more alternative strategies than could be reasonably attempted on a practical field scale (Eadie and Maxwell, 1975). To take account of the wide variation in quality of hill herbage and the ability of the hill sheep to select preferred components, herbage was conceptually classified in terms of its digestibility; herbage growth was defined as the addition of new material to the highest digestibility class and deterioration as the movement of existing material down the digestibility range. These procedures allowed a description of herbage to be maintained throughout an annual cycle with varying growth and deterioration rates. The selection of herbage assumed that material of high digestibility was preferred but that the degree of selection achieved would be limited by grazing pressure and herbage structure, as represented by the digestibility classes in the vertical and horizontal planes. Intake was restricted according to the quality of the diet selected and the size of the grazing sheep.

Essentially, then, the model was concerned with herbage growth, diet selection and the maintenance and live-weight change of wether sheep (Sibbald, Maxwell and Eadie, 1979). Before it can be used to examine effects on animal production it will be necessary to add the components of pregnancy and lactation (Chapter 4). Although it was based largely on theoretical concepts of herbage growth, selective grazing and herbage consumption, the output of the model was in reasonable agreement with the field data of Eadie (1967b) and supported the generally accepted conclusion that it is the low quality of the herbage that sheep are able to select during the winter which determines the annual stocking rate sustained in traditional set-stocked hill sheep systems (Sibbald *et al*, 1979).

From 1973 the model was modified and developed by Vine (1977) to examine the effects of grazing management on herbage production and utilisation in intensively managed grassland. Use of this model to investigate the grazing process in a single-species sward clearly indicated a paucity of knowledge in certain crucial components. These were (a) the distribution and rate of change in the concentration of digestible energy within the sward canopy, (b) the regrowth of herbage following grazing, and (c) diet selection.

The formulation of the model demonstrated that the tiller can be used as the basic unit of account. The experimental results show that large changes in digestibility occur as leaves age from one leaf position to the next. The intake of digestible herbage by a grazing animal could be more closely monitored if changes in weight, digestibility and location of leaves as they aged were represented as continuous variables. To this end Vine proposed the concept of a 'standard leaf interval' scale, subdivided into classes which describe the age of a leaf, not in chronological terms, but in terms of its change in position on the tiller. Using this concept, the amount of herbage can be represented as a distribution of dry matter weight against a continuous scale of leaf position. The rate at which material passes through this scale is determined by the rate of leaf turnover and the concept provides a common basis for representing the growth of herbage, its structure, and the removal of material from it.

Any conceptual framework or quantitative model which purports to simulate grazing must represent the sward in a way which will be relevant and sensitive to the process and mechanics of grazing, and will be amenable to an adequate representation of herbage growth and senescence. In its discussions the Grazing Studies Group has therefore been concerned, in part, with the compatibility of the various levels of biological organisation, and hence the detailed experimental measurements, that are necessary to meet these requirements. There are similarities between the sets of sward characteristics likely to be involved in determining herbage growth on the one hand and diet selection and herbage intake on the other, since both involve a description of the distribution of plant parts within the crop, though they may differ in the apparent degree of detail required. Following the work of Vine (1977), the individual tiller has been adopted as the basic unit which can be modelled to take account of sward characteristics which have to be continuously updated through time in response to patterns of growth, senescence and defoliation. As it is now possible to measure the bite size, number of bites and the quality of material in a bite of a grazing animal another possibility arises, which is to consider the pasture in terms of a bite size/site classification. Milne (1977b) applied this approach to the results of concurrent heather grazing studies (Grant *et al*, 1978; Milne *et al*,

1979), and examined the factors that influence the selection of particular parts of the heather plant.

Other alternative approaches may emerge. Ultimately, the level of biological organisation that can be modelled reasonably will be a function of the ability to measure and describe the various sward characteristics. Although it is apparent that some experimental data on herbage intake can be simply explained and modelled in terms of the weight, height or allowance of herbage, it has not been possible to develop a unifying concept from the evidence available (Hodgson, 1977). Similarly, no all-embracing hypothesis has yet emerged on which herbage selection can be based. The way in which we think we will need to represent pasture merely reflects the concepts we have of how selection by the animal of particular plant parts might operate; further meaningful progress is entirely dependent upon experimentation. So far, modelling has provided a means of isolating some of the issues which require further examination in this regard.

There still remains the issue of how best to deal with herbage growth. It can be argued that, if the effect of defoliation through grazing on regrowth of pasture is to be adequately represented, pasture growth will have to be derived from a model of carbon exchange influenced by moisture status and the supply of available nutrients. However, a major difficulty arises in modelling the partitioning of assimilated material into individual tiller leaves, stem inflorescence and roots of an individual tiller. This requires further consideration. The factors influencing the development or loss of tillers in the swards are also imperfectly understood. While recognising the importance of developing a model of tiller dynamics relative to changes in environment and management (Hodgson and Wade, 1979) the Grazing Studies Group concluded that at present the paucity of relevant data and lack of understanding of underlying physiological processes probably precludes all but a mechanistic or empirical approach to modelling tiller development.

In summary, the modelling of sward-animal interactions at HFRO has led to a much sharper appreciation of the issues which require to be resolved as a means of developing more effective systems of grazing management. The usefulness of modelling as a framework for communication amongst those concerned with pasture production, animal production and systems research may be judged to some extent on the basis of the research programme that has developed in recent years.

Grazing studies on sown swards

The increasing interest in the consequences of variations in grazing management upon herbage and animal production from enclosed upland pastures and improved areas on hill farms led in 1972 to official approval

Improved and
indigenous
pastures
at House
o' Muir

Effect of utilisation on
dead/green ratio in pasture

for work in the context of upland farming. This resulted in the development of research projects on upland sheep and beef cattle, which are dealt with in Chapters 7 and 8 respectively, and to a rapid expansion in studies on sown swards. Much of the initial detailed work was carried out on predominantly perennial ryegrass *(Lolium perenne)* swards, although it was recognised that most upland swards would be mixed, and should sustain a substantial clover *(Trifolium repens)* content (Chapter 3).

Interest in this area arose during the second phase of the grass-heather experiment when treatments were managed so that the herbage weights on the grass sub-plots were maintained at different levels (400, 950 or 1500 kg DM/ha). Estimates of tissue production and loss on the grass sub-plots were made from measurements of rates of leaf extension and senescence per tiller, weight per unit leaf length of immature and mature green leaf and numbers of tillers per unit area. Differences were found in production on a per-tiller basis but these were offset by adjustment of tiller densities. The results indicated that management effects on herbage production over the period from late May to early August in the dry summer of 1976 were negligible compared with fluctuations associated with climatic variables, particularly moisture stress.

At the same time Vine (1977) was carrying out detailed studies into rates of leaf production and loss, the relationship between leaf maturity and digestibility, and the distribution of digestible nutrients within a perennial ryegrass sward. The results of these two sets of studies, together with evidence from studies of leaf and tiller dynamics in grazing experiments, suggest that adjustments in production per tiller, together with fluctuations in tiller density, act as a homeostatic mechanism whereby net dry matter production per unit area will tend, within limits, towards a norm for a given soil, climate and sward. This plasticity of response has been examined in more detail in a trial involving the selective removal of individual leaves in mini-swards in a greenhouse and in current grazing experiments where herbage growth rates are being measured in swards in which the amplitude and time scale of fluctuations in herbage weight are varied under controlled conditions. These studies are linked closely to observations on the patterns of defoliation of individual leaves and tillers in grazed swards.

King, Lamb and McGregor (1979) have recently studied the relationship between leaf area and net photosynthesis in both cut and continuously stocked swards of perennial ryegrass. Adaptive changes in sward structure and photosynthetic efficiency tended to modify the influence of repeated defoliation on carbon assimilation, although leaf area index was still the principal factor influencing this (Fig. 12). There is, as yet, no information on the correlation between estimates of carbon balance and of rates of herbage growth in the sward, but this relationship is being studied in a current grazing experiment involving the measurement of carbon exchange

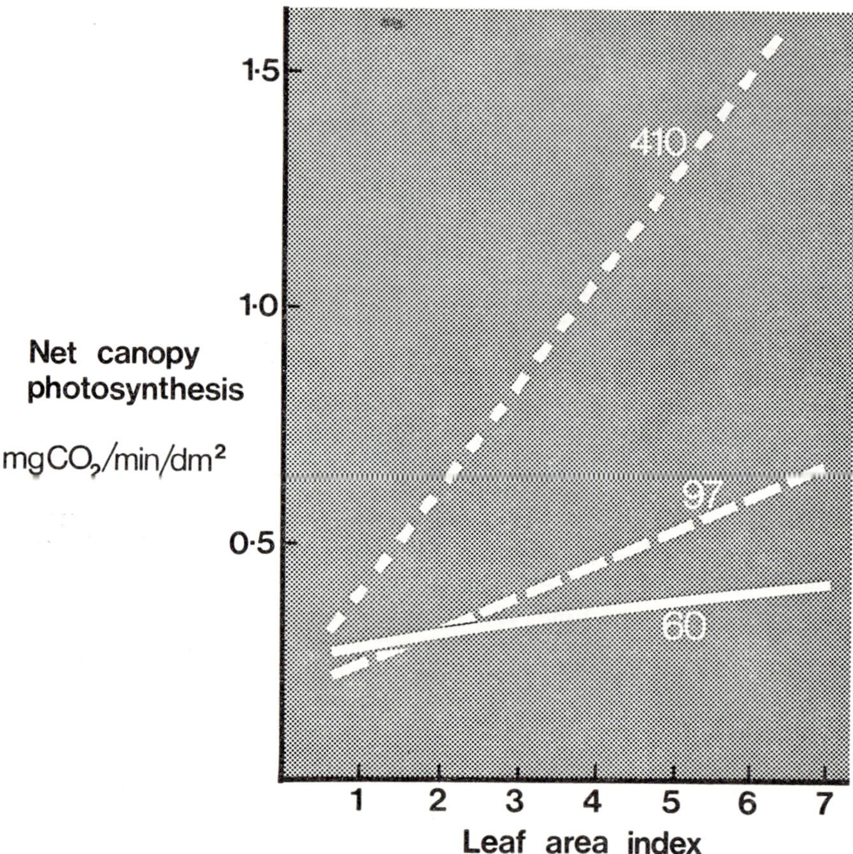

Fig. 12   The relationship between net canopy photosynthesis and leaf area index in continuously stocked swards, measured at three light intensities (60, 97 and 410 watts/m² visible light).

and partitioning in the sward, and rates of tissue turnover in terms of the production, consumption and death of leaves and tillers.

Parallel experiments with grazing sheep have shown that daily herbage intake is positively related to the weight, height and density of herbage in the sward, and suggest that the relationship between daily herbage allowance and intake is mediated via these sward variables (Hodgson and Milne, 1978).   These responses appear to be a function of variations in the mechanics of the grazing process (i.e. grazing time, the rate of biting during grazing, and bite size).   As a result, much attention is now devoted to measurements of these behavioural variables in grazing studies, making increasing use of automatic recording equipment, and to short-term measurements of rates of herbage intake using animals fistulated at the oesophagus. Since most livestock farms carry stocks of both cattle and sheep, there is increasing interest in comparative studies on these two species.

Studies on the management of white clover began in 1976 with a pot study on the effect of pattern of defoliation of individual clover plants (King, Lamb and McGregor, 1978) which showed that partial defoliation affected various parts of the plant differently, according to which leaves were removed. Since then, studies have been initiated on the influence of clover content and sward structure on selective grazing and herbage intake by sheep grazing mixed grass/clover swards, and on the influence of grazing management on the clover content and herbage dynamics of mixed swards, this latter study involving some collaboration with the group working on nitrogen fixation (Chapter 3). The trend to increasing emphasis on work with mixed grass/clover swards is likely to continue.

In recent years herbage intake measurements have also been made regularly in production studies with sheep and cattle (Chapters 7 and 8). In the sheep studies particular attention has been devoted to the influence of sward conditions on herbage intake during lactation and in the breeding season, and there has been a major interest in the interaction between the effects of sward conditions and the level of concentrate supplementation on herbage intake, milk yield and lamb growth in early lactation (Milne, Maxwell and Souter, 1979). In the beef cattle studies attention has been concentrated on the interaction between the effects of winter nutrition and sward conditions on the herbage intake of spring calving cows (Hodgson, Peart, Russel, Whitelaw and Macdonald, 1978) and in the early development of herbage intake in their calves. This work is now being expanded to allow a more detailed examination of the relationships between the effects of the previous nutritional history of the animal and its current productive state, sward conditions and grazing management upon herbage intake. The objective of these studies is to put decisions on grazing management and supplementary feeding within animal production systems on a sounder footing than is possible at present.

Grazing studies on indigenous hill swards

The use of information on seasonal cycles of nutrient intake by grazing sheep to design practical systems of management to make the most effective use of *Agrostis-Festuca* and *Calluna* communities at Sourhope and Glensaugh respectively has already been described. Current grazing studies on hill plant communities are intended to increase the scope of this information, and to provide comparative data for cattle and sheep, in order to allow objective decisions to be made about appropriate managements for a range of hill conditions, including possible combinations of vegetation types and animal species.

A phased, interdepartmental study started in 1977. Information is sought on (a) seasonal variations in the potential nutrient intake achieved from specified plant communities, (b) differences between cattle and sheep in nutrient intake and diet selection in relation to detailed measurements of sward structure and botanical composition, and (c) the impact of grazing management on the communities themselves. Pastures representative of the main categories of hill vegetation are under study, including *Agrostis-Festuca* acid grassland, *Nardus*-dominant grass heath, *Molinia* grassland, heather moor and blanket bog. Measurements are also being made on a ryegrass community to act as a link with other work on sown pasture.

Initially, grazing pressures are being kept deliberately light, approximating to those of traditional systems. In the second stage of the work, grazing management will be manipulated within the context of possible systems and the impact on floristic composition, herbage production and nutrient intake by grazing animals measured. The study includes an evaluation of the role of cattle as possible improvers of vegetation, and should provide an assessment of the validity of the suggestion (Darling, 1955; McVean and Lockie, 1969) that a widening of the ratio of sheep to cattle has been responsible for undesirable vegetation changes in hill areas over the last two centuries.

Prospect

The recent work at Glensaugh on the utilisation of heather communities repeated in essence the approach used in the original studies on grass and grass heath communities at Sourhope. Although the solutions proposed differ substantially, the commitment to achieving improvement in output through an understanding of the opportunities for manipulating the relationships between sward and animal, following an analysis of the underlying limits to production, was essentially the same. The current studies on a range of plant communities developed logically from this earlier work. They are likely to follow the same cycle of activity, giving rise in their turn to further systems development projects, and at the same time providing a basis for assessing the appropriate balance between sheep and cattle on hill plant communities.

Studies on the management of improved swards are a relatively new departure for the Organisation, reflecting, as already indicated, the acknowledgement of a remit for work in the context of upland farming. The Grazing Studies Group has made a valuable contribution to the development of this work, the formal framework for the discussion of topics of mutual concern by individual scientists with a range of specialist interests having already given rise to several projects concerning tissue dynamics and patterns of defoliation in grazed swards.

There is repeated evidence throughout this chapter of the developing links between the plant and animal studies which have formed the core of the Organisation's research programme and the work on grazing ecology. For instance, many of the programmes on plant and animal nutrition (Chapters 3 and 4) now involve detailed measurements of patterns of plant growth and/or consumption under grazing conditions, evidence of the recognition that the relationships between animal and sward can be major determinants of herbage and animal production under pastoral management. In turn, grazing ecology studies may themselves be conditioned by information, often from plot experiments or feeding trials, on the limits to manipulation by grazing management which are likely to be of practical relevance.

It is clear that attempts to achieve a better understanding of the factors governing the response of the sward to defoliation by grazing animals, and the response of the animal to variations in sward condition, are likely to form a major component of our grazing studies in both hill and upland environments for some time to come. In this sense the programme of work on indigenous and sown swards are complementary, and should lead to a better understanding of the basis on which to plan the most effective integration of the varied plant and animal resources which characterise most of our hill and upland farms.

References

ARMSTRONG, D G and THOMAS, B (1953). The nutritive value of *Calluna vulgaris*. 2. A preliminary study of digestibility. *J Agric Sci, Camb*. 43, 223-228.

DAFS/NCC (1977). *A guide to good muirburn practice*. Muirburn Working Party, Edinburgh, HMSO.

DARLING, F F (1955). *West Highland Survey: An essay in human ecology*. Oxford Univ Press.

DOAS (1953). *Hill Farm Research Second Report*. The Scottish Hill Farm Research Committee. Edinburgh, HMSO.

GB PARL (1972) HC 511-i. Select Committee on Scottish Affairs Session 1971-2. *Land resource use in Scotland*. Vol 1 Report and proceedings of the Committee, London HMSO, 90-91.

McVEAN, D N and LOCKIE, J D (1969). *Ecology and land use in upland Scotland*. Edinburgh Univ Press.

TILLEY, J M A and TERRY, R A (1963). A two-stage technique for the *in vitro* digestion of forage crops. *J Br Grassld Soc*. 18, 104-111.

6

Systems research

Contributors: R H Armstrong, D C Currie, J Eadie, A R Sibbald, A Whitelaw
Co-ordinator: T J Maxwell

Introduction

Analytical research is concerned with obtaining an understanding of the biological components of production. It reduces a system of production to a set of simpler components, for example reproduction, pregnancy, lactation and lamb growth, in order to understand and quantify the factors affecting them. In such research interaction between variables is usually reduced as far as possible. An understanding of the separate components is seen as a means of gaining an insight into the whole system of production.

In a production system, however, many components interact with each other as well as being determinants of the properties of the whole. Decisions made at any particular time affect the system not only at that point but throughout the production cycle. A full understanding of a system therefore cannot be obtained solely by studies of its components in isolation.

Systems research is concerned with putting components together and with evaluating the systems of production which result from such syntheses. It can also involve experimentation within the system which permits one or more of the components to be varied and controlled. In association with analytical research, this approach provides an understanding of how to allocate and manage resources to achieve particular objectives. In agriculture the objective of systems research is to improve productivity in an appropriate way.

The hill sheep systems development programme in the Organisation began in 1968 and has been concerned with the testing of systems of production very different from those traditionally practised. An explanation of the syntheses of these improved systems in terms of the various components of production and the method by which they were tested is presented in this chapter. An account of the various development projects is given and some of the implications of the results to the analytical research programme are discussed. Much of the material presented is drawn from an initial outline and discussion of the systems development work in the Organisation by Eadie (1970a) and from a later review by Eadie and Maxwell (1975). The work on upland sheep systems will be described in Chapter 7.

The hill sheep systems development programme has involved much of the land and many of the staff at each of the Organisation's research stations. Dr R H Armstrong, officer-in-charge at Sourhope, and D C Currie, officer-in-charge at Lephinmore, have both played prominent roles in the development of the programme.

Limited resources of land and staff and the need to investigate more alternatives in systems input and formulation than is possible in the field has made it necessary to investigate other ways of studying systems. Thus, for the last six years mathematical models of some of the biological components of animal production from hill and upland pastures have been constructed to facilitate the eventual modelling of 'sheep production systems. The component models are reported in the relevant chapters of this report.

Hill and upland agriculture have to be considered in relation to other competing land uses, in particular the forestry industry. Both agricultural and forestry interests have regarded land allocation decisions as lacking objectivity and a model has been evolved which attempts to provide a logical framework for decision making (Maxwell, Sibbald and Eadie, 1979).

Hill sheep production systems

Hill sheep production systems are concerned with the conversion of hill and mountain pastures to saleable sheep products. These pastures, classified in the agricultural statistics as rough grazings, are uncultivated and are often at elevations above 250 m. They encompass a wide range of soil and climatic conditions and, consequently, include a wide variety of vegetation types (Chapter 2).

Traditionally, hill sheep are set-stocked in a free-grazing system and are expected to obtain most of their nutrient requirements from grazed pasture throughout the year (Chapter 1). The flocks are self-replenishing and output is in the form of weaned lambs, wool, and ewes culled for age or other reasons. As a consequence of low stocking rates and low levels of individual sheep performance, output per unit area, although varying greatly from one hill region to another, is low. Average weaning percentages are in the region of 80 and range from below 60 to over 100; lamb live weights at weaning range from about 20 kg to above 30 kg.

The biological and economic components of production

By 1967 a good deal of information had been accumulated from the Organisation's research programme on the various components of production, and interest in examining this in relation to the development of improved systems of production was increasing. The first step was to examine the knowledge available about the traditional system and its components to identify the possibilities that existed for overcoming the inherent problems.

The vegetation found on hill farms and the limitations to its production and utilisation are outlined in Chapter 2. The means of improving the utilisation and quality of some indigenous hill grazings are indicated in Chapter 5. The factors and procedures relating to the replacement of indigenous hill

1969   1978

Hairney Law

Mid Hill

Pastures used
in hill sheep
systems research

grazings and the establishment and maintenance of sown pastures are dealt with in Chapter 3.

Individual animal performance and hill sheep production are functions of the nutrition of the ewe (Chapter 4) and the system of grazing management (Chapter 5). Research up to 1968, when the systems development programme began, clearly indicated that the annual cycle of nutrition afforded by the traditional systems of hill sheep management was a significant constraint on animal performance (Russel, 1971).

The economic circumstances of the industry constrained the range of the possible improvements to systems of hill sheep production. For many years up to the mid-fifties hill sheep farming was characterised by low levels of output which, in general, were matched by a correspondingly low cost structure. In the following decade, Duthie (1968) found that the average selling price of both store lambs and cast ewes in the east of Scotland hardly changed and that, despite the change in the value of the pound, net profit remained static. During this time Government support rose from 60-70% to 130-190% of the net profit. From 1964-69 both fixed and variable costs increased and profits declined in real terms by about 70% (Lord and Blyth, 1973) despite continued Government support. Another feature of the industry which had become increasingly apparent in the late sixties was a trend to reduce fixed costs by economising on labour. It was clear that this trend could not continue indefinitely since many hill farms were already one- or two-man units.

While output remained low the size of business generated from many hill farms was too small to remain viable in the economic circumstances of the late sixties and early seventies (Allen, 1973). Increasing the size and scope of a holding by renting additional land may have helped if it was within reasonable distance of the original holding. The capital involved in the acquisition of more land precluded most farmers from increasing the size of their holding by land purchase (Harkins, 1970). The overall impact of amalgamations on the economics of hill farming was, therefore, likely to be small.

This was the state of the industry when the development programme began. The economic aspects of the programme was therefore regarded as being very important. It seemed that the most satisfactory approach would be to create a system of production which could stimulate output to a level which would not only cover any additional costs but would generate sufficient increased income to secure the further development and future economic viability of hill farms.

Synthesis

Biological possibilities

The understanding and knowledge of the biological components of production suggested the need for a very different system to that traditionally practised. The biological possibilities were, of course, subject to non-biological constraints, for example physical, operational and economic factors. The need to improve ewe nutrition throughout the annual cycle of production, and the possibilities which existed to achieve this from pasture, clearly indicated changes in grazing management towards improving ingested herbage quality and herbage untilisation.

Higher levels of utilisation had been shown to eliminate much of the accumulated senescent herbage and to give rise to diets of higher quality on a limited range of hill pastures types, notably *Agrostis-Festuca* which constitutes some 15-40% of many grassy hills (Eadie and Black, 1968). On other pasture types such as *Nardus-* and *Molinia*-dominant grass heaths, better utilisation gives rise to a substantial short-term nutritional penalty and to a much less productive and nutritionally poorer pasture. On the vegetation types found on blanket peat, the more valuable species are likely to be grazed at intensities well below those consistent with high levels of utilisation, and evidence suggests that much higher levels of pasture utilisation than those currently achieved can be sustained, although improvement in pasture quality will be slight (Grant and Campbell, 1978 and unpublished).

Most hill grazings are vegetationally heterogeneous. On the basis of the evidence presented and discussed in Chapters 2 and 5 on the levels of pasture utilisation and quality in traditional grazing systems, it was argued that a form of grazing control which would take account of these various classes of vegetation was a necessary prerequisite to higher animal output from indigenous pasture. It was also argued that although indigenous pastures could be replaced by sown pastures, there was little point in increasing pasture production on part of a grazing unless steps were taken to ensure its efficient utilisation; this also implied grazing control. A distinction was made between those pasture types, sown or indigenous, which could be managed to provide feed of much improved quality, and the poorer quality indigenous hill pastures from which only better utilisation could be expected.

Improvement in animal performance required improved levels of nutrition from pasture, particularly during lactation and the pre-mating and mating period. One objective of the new system was to achieve better pasture utilisation, especially of those pasture types capable of providing improved nutrition during these critical periods. This objective also implied a potential for increasing stock numbers.

In anticipation of the effects of improvements in summer and autumn nutrition, it was necessary to provide levels of winter nutrition which would take account of improved conception rates and increases in stocking rate. Evidence suggested that acceptable levels could only be achieved by the use of cereal/protein concentrate during the last six to eight weeks of pregnancy. Ewes would be dependent on indigenous pasture for much of the rest of the winter. It was recognised, however, that an alternative approach would be to remove pasture dependence at this time by off-wintering thus allowing a much closer relationship between stocking rate and summer herbage production to be achieved. This would give rise to more efficient herbage utilisation but would also incur substantial increases in the cost of winter feeding.

Non-biological constraints

The physical constraints referred to earlier concern pasture improvement. This, in its various forms, requires safe access for vehicles and materials. The topography of much hill country prevents access to large areas of land and renders reseeding impossible on substantial parts of many grazings.

Many of the operational constraints concern the use of labour. Apart from the fact that such costs are a significantly greater proportion of total costs in hill sheep farming than in other types of livestock production, labour is often in limited supply. Improved systems must therefore attempt to make more effective use of labour, for example fencing provided to achieve the grazing control required to satisfy the biological objectives also allows for more efficient labour use.

The economic implications of the costs of pasture improvement, supplementary feeding, and increased stock numbers, have to be considered in relation to the increases in output that can be expected from an improved system. In the late sixties the continued downward trend in profits and the seeming lack of any indication that this trend would be reversed severely restricted the availability of capital for investment and indeed this appeared to be a major impediment to further progress. An appropriate means of economic appraisal developed at the time was to assess the return on the marginal capital investment. Harkins (1968) produced a comparatively simple procedure to calculate the increase in output required to break even at various levels of capital investment. This was based on a break-even budget technique and calculated increase in output in terms of the possible combinations of stocking rate increase and improvement in sheep performance. The proportion of the capital invested at any one time, the rate of flock expansion and the rate of improvement in individual performance required further consideration in respect of their effects on cash flow and liquidity (Maxwell, Eadie and Sibbald, 1973b).

Economic arguments also led to the conclusion that summer nutrition during the periods of lactation, lamb growth and the recovery of ewe body condition should not be improved by supplementary feeding.

The economic constraint on off-wintering derives from the need to purchase a very high proportion of winter feed. To justify this expenditure it was necessary to establish the extent to which stocking rates could be increased and animal performance improved.

There were two possible systems of production suggested — a year-round grazing system and an off-wintering system.

Year-round grazing systems

Two kinds of pastoral resources are identified. One, of high nutritional quality, is used during lactation and lamb growth and again, after a mid-season rest, in the period before and during mating, and the other, the poorer quality indigenous hill pasture, is used throughout the winter and again between weaning and pre-mating. Lambs are removed from the system at weaning, as occurs traditionally; ewe hoggs are removed to an enclosed grazing during the winter, or off-wintered, and in the following summer grazed on the unimproved hill. Other management changes which are associated with this system of controlled grazing are the provision of better management of rams which are raddled at mating, the provision of semi-permanent winter feeding points and lambing in the enclosed areas.

Off-wintering systems

Stocking rates in traditional systems are set at levels determined by the need to maintain certain minimum levels of winter nutrition from grazed pasture (Eadie, 1970a). Systems based on off-wintering offer possibilities of substantial increases in output as a consequence of the removal of this major constraint.

The understanding of winter nutrition and its effect on animal performance suggested that in-wintering of itself would have little effect on increasing output. This was supported by the limited evidence available from in-wintering studies at that time. Even if the capital cost of a wintering house could be avoided by off-wintering sheep on an area of pasture or some form of hard standing, increases in output necessary to justify the expenditures involved are likely to be in excess of 100%. However, the introduction

Table 5:   Summary of management plan

Phase of cycle	Resource			Feeding		Objective
	Hill	Improved pastures	Off-wintering area or Inwintering house			
Pre-mating (Oct-Nov)	Ewe hoggs	Ewes and gimmers	—	—		Improve weight and condition
Mating (Nov-Dec)	—	Ewes and gimmers	—	—		Maintain weight
			Ewe hoggs (until April)	Hay and concentrates until April		Achieve weight gain of 5-7 kg by April
Mid-pregnancy (Jan-Feb)	Outwintered ewes and gimmers	Pasture rested	Inwintered ewes and gimmers	*Outwintered* Storm feeding only	*Inwintered* Hay/sugar beet pulp and cereal/ protein supplement	Weight loss not more than than 5 kg
Late pregnancy (Mar-Apl)	Early and late lambing ewes and lean ewes and gimmers — separated where possible	Pasture rested	Early and late lambing ewes and lean ewes and gimmers separated	Introduce and gradually increase cereal/protein supplement	Gradually increase supplement	Weight increase of 6-8 kg (Plasma 3-OHB 0.8 mmol/l)
Lactation (Apl-Aug)	Hoggs and dry ewes	Ewes and gimmers: priority to ewes with twins and to gimmers	—	Supplement continued until adequate herbage available — usually 3 weeks		High level of milk production, high lamb growth rate and some ewe weight recovery
Recovery (Aug-Sep)	Ewes, gimmers and hoggs	Pasture rested	—	—		Improve weight and condition

of an area of improved pasture integrated with the open hill as outlined for the year-round grazing systems, which would have the effect of improving individual animal performance, suggested the possibility that the increases in stocking rate necessary to reach an economic break-even point could be reduced. Inwintering was chosen as the method of off-wintering because of the control it allowed over winter nutrition and its mitigating effects on between-year variation in winter weather conditions.

In the management plan ewes were to be in-wintered from the beginning of January, after the completion of mating, up to lambing. Thereafter they could be managed in the same way as in the year-round grazing system. The two systems that were proposed are summarised in Table 5.

Evaluation of systems of hill sheep production

The objectives of evaluating systems of hill sheep production were, firstly, to establish from a management point of view that they could be operated satisfactorily; secondly, to measure the responses of total output and individual animal performance to specified changes in inputs over a range of stocking rates; thirdly, to provide biological data for quantification of the relative contribution that each component of the system made to the output of the whole; fourthly, to assess the input/output relationships in economic terms.

The process of evaluation is not concerned with achieving a single acceptable and satisfactory level of output and animal performance to demonstrate the success of a system but with the need to establish where limits lie. For example, because stocking rate has an important effect it has to be increased to levels which depress animal performance significantly before any further inputs are made.

In considering the method by which systems could be adequately tested and evaluated in the field several important factors had to be taken into account.

Behaviour and vegetational diversity

Traditional systems rely on an important behavioural attribute of hill ewes which display a distinct diurnal movement between the lower and higher ground of the grazing and exhibit a well defined territorial beheviour known to be nutritionally significant (Hunter, 1962). The proposed system of controlled grazing management could grossly interfere with traditional behaviour.

Another important feature of hill sheep farming is the considerable variability of its land resources. The significance of different proportions and distributions of vegetation on the year-round nutrition of the ewe on any unit has already been emphasised, but its effect on number of stock carried and levels of output achieved are also important in the evaluation of systems of production. The positive correlation of size of unit and vegetational diversity is likely to be especially important where the improved pasture is created by grazing management alone or with soil improvement, i.e. where no replacement of indigenous vegetation by sown species is contemplated.

In considering the implication of the effect of vegetation and behaviour on production and performance, and since it was anticipated that many of the problems would be associated with size and scale of operation, it was thought necessary to establish system studies which use areas and numbers of animals approaching those found in practice. The possibility of finding two sufficiently similar hill grazings of adequate size within the boundary of one property is very remote and consequently the classical experimental approach of treatment, control and replication could not be used in this work.

An alternative approach was to define rigorously the resources, objectives, management decisions and inputs (for example Table 5) and to measure, in conjunction with an adequate programme of biological and economic monitoring, the levels of total output and animal peformance achieved. The extent to which biological monitoring could be carried out had to be viewed against the available staff resources and the importance of testing the systems over as wide a range of environments as possible.

Stocking rate

The implication of changes in stocking rate in the study of systems of hill sheep production has been discussed by Eadie (1970a).

The level of output achieved from a system will be a function of its stocking rate and level of individual performance for a given level of inputs, but individual performance and stocking rate are often inversely related. A system can be assumed to have reached stability when the annual ewe live weight cycles in a characteristic way over a characteristic range, but the cycle typifies events at only one particular stocking rate. If more efficient use of existing or enhanced pasture production is to be achieved increases in stocking rate are required. These are made over a range sufficient to identify, firstly, the point at which total output reaches a maximum and, secondly, the point at which individual animal performance declines significantly. A diagrammatic representation of this procedure is given in Fig. 13.

In summary, the method adopted has measured the responses to a given·set of inputs and associated management procedures on a given unit over time and over a range of stocking rates which have encompassed the

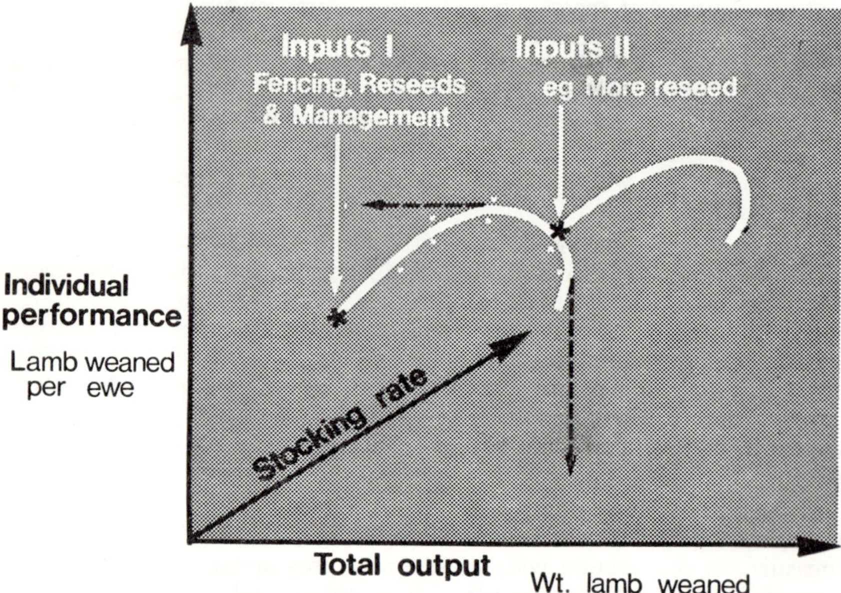

Fig. 13 Theoretical relationships amongst stocking rate, individual ewe performance and total flock output for given inputs.

maximum level of total output and individual animal output achieved. The procedure has been less questionable than it might have been because, as was predicted, output responses (total lamb output and output per ewe and per unit area) have been sufficiently great (not less than 50%). The significance of the changes in performance can also be related to the data collected from the programme of biological monitoring.

Biological monitoring

A considerable emphasis was placed on the monitoring of ewe and lamb live weights throughout the annual cycle in association with the other parameters of animal performance, for example mortality, lambing percentages, etc. The data have been collated and summarised for each of the projects (Sibbald, unpublished). They have provided some understanding of the interactions amongst the various components of production, and have served as a backcloth to some of the analytical research topics which currently form part of the Organisation's research programme. The data are used as a basis for making management decisions and are of vital importance in interpreting the changes in output and individual animal performance that take place in the various studies.

Vegetation and soils have been monitored at intervals during the life of the systems studies; the implication of some of the changes that have taken place are discussed later in this chapter. Veterinary monitoring was considered to be important and is also dealt with later.

Economic evaluation

All aspects of capital investment, annual expenditures and income have been recorded for each of the systems studies. Gross margin and cash flow analyses were prepared, and returns on capital investment calculated.

In planning the development of a system and assessing the possible cash flows arising out of an investment, a considerable number of calculations must be made, particularly where comparisons between alternative policies are required. These calculations involve different rates of improvement in animal performance and stock number increase, and different timing of capital investment. An econometric model (Maxwell, Eadie and Sibbald, 1973a; Sibbald and Maxwell, unpublished) was designed to deal with these alternatives.

Year-round grazing systems programme

Since 1968 five year-round grazing studies have been initiated in three broadly different soil, vegetation and climatic environments.

Hairney Law/Auchope (Sourhope)

The Hairney Law/Auchope year-round grazing study at Sourhope has been reported in detail (Armstrong, Eadie and Maxwell, 1978). This study was based on indigenous hill vegetation improved by grazing control (Chapter 5). Initial capital inputs were restricted to fencing and to the retention of extra stock ewe lambs to allow a build up of stock numbers. The area chosen for the study extended to 283 ha, ranging from 250-490 m, with 35-40% of the area being *Agrostis-Festuca* pasture, the remainder a *Nardus-* and *Molinia*-dominant grass heath. Originally, the unit carried 398 Cheviot ewes and 25 suckler cows.

Phase 1: 1968/69-1972/73. Five areas of *Agrostis-Festuca*, each of approximately 20 ha, were enclosed and an initial heavy stocking with cattle and

Table 6:   Flock production data (Hairney Law/Auchope)

	Base data	Phase 1					Phase 2				
	5 yr average	68/69	69/70	70/71	71/72	72/73	73/74	74/75	75/76	76/77	77/78
Stock numbers (ewes and gimmers)	387	398	451	518	528	573	600	601	620	623	623
Weaning percentage	90.6	84.7	86.7	102.5	104.7	99.5	91.5	102.7	108.5	106.9	105.1
Weight of lamb weaned total (kg)	7924	7786	9188	14,177	14,046	14,193	14,329	16,042	17,902	17,596	16,470
per ewe (kg)	20.5	19.6	20.4	27.4	26.6	24.8	23.9	26.7	28.9	28.3	26.4
Total weight of wool (kg)	869	850	1017	1253	1369	1561	1454	1535	1543	1503	1523

dry sheep was used to remove as much as possible of the previously ungrazed herbage. These areas of improved pasture were subsequently used during early and mid-lactation and again during the pre-mating and mating period. Lambing took place in fields which initially were outwith but adjoined the area chosen for the study. The cows grazed on the unit from May to December each year.

The flock production data are given in Table 6. Ewe numbers increased from 398 to 573. Total lamb output increased from 7924 kg to 14,193 kg and lamb output per ewe reached a peak of 27.4 kg in 1970/71 at a stock number of 518 (Fig. 14); capital investment, net of grants, during

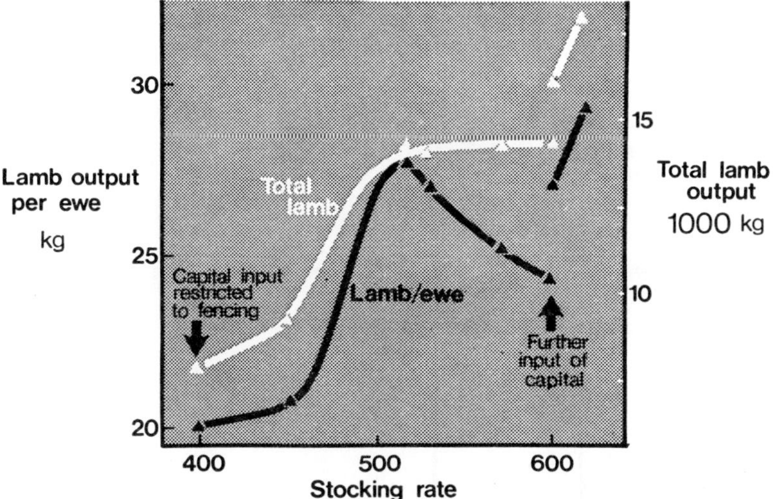

Fig. 14　Hairney Law/Auchope — relationships amongst stocking rate, total weight of lamb weaned and weight of lamb weaned per ewe mated.

this period was £689 and flock gross margin increased from £1445 in 1968/69 to £5607 in 1972/73. The cash flows from the investment gave rise to a cumulative balance of £1395 in 1972/73.

The pre-mating live weights of the breeding flock, and live-weight gains or losses are given in Table 7. During 1972/73 the ewe live-weight data indicated that performance had reached a maximum in relation to ewe numbers. Live-weight losses during the mating period increased substantially, and live-weight recovery during and after lactation was reduced, and these, together with a further substantial loss before the next mating, resulted in an appreciable decrease in weaning percentage in 1973/74.

Table 7:   Ewe pre-mating live weights (kg), and live-weight changes (kg) throughout the year (Hairney Law/Auchope)

Production Year	Pre-mating live weight	Pre-mating to Post-mating 16/11-5/1*	Midwinter 5/1-12/3	Pre-lamb feeding 12/3-10/4	Pre-lamb to marking 10/4-24/5	Marking to weaning 24/5-8/8	Weaning to early October 8/8-2/10	October to Pre-mating 2/10-16/11
1968/69	50.9	−3.5	−3.2	+3.3	−3.2	+4.9	+4.6	−0.4
1969/70	53.2	−3.5	−1.8	+3.1	−5.0	+4.8	+3.3	+2.1
1970/71	56.0	−1.6	−1.1	+2.7	−3.6	+2.1	+3.1	−0.1
1971/72	57.7	−1.6	−1.1	+1.1	−3.0	+2.4	+4.0	−0.2
1972/73	59.3	−7.2	+2.8	+4.0	−8.7	+3.6	+3.4	−3.5
1973/74	53.7	−3.2	−0.7	+2.3	−2.8	+5.7	+1.8	−1.0
1974/75	55.7	−2.4	−0.9	+3.0	−5.6	+5.1	+2.6	−0.4
1975/76	58.0	−1.8	−2.5	+2.5	−3.9	+4.0	−1.0	1.7
1976/77	53.6	−1.8	−0.5	+4.0	−6.5	+7.5	+1.4	0.0
1977/78	57.7							

* All dates shown are average dates over 9-year period, when each specific weighing carried out.

Phase 2: 1973/74-1976/77. Further inputs of improved pasture were made and financed out of the cumulative balance already achieved from the earlier investments. The existing enclosed *Agrostis-Festuca* paddocks received a dressing of ground magnesium limestone and ground mineral phosphate. Some of these areas were also sprayed to eradicate bracken, 'spike bar' rotovated and oversown with a grass and clover mixture. In addition the areas of reseed that had been used only for lambing up to 1973 were included on an all-the-year-round basis from 1974. During 1977 5.5 ha were rotovated and completely reseeded with a grass/clover mixture. Since 1974 a further £3034 has been invested in the Hairney Law/Auchope unit. Ewe numbers have increased but only by 48 (8%) since 1973. Total output of lamb has increased to 17,596 kg in 1977, output per ewe having increased to 28.3 kg. The cumulative balance derived from the investments since 1968 totals £10,590 to 1977.

Alderhope (Sourhope)

In 1972 a further study was initiated on the Alderhope unit at Sourhope to examine the impact of land improvement brought about by reseeding. Interest centred on establishing the rates of improvement in animal performance and stock carrying capacity that were possible, and whether they were adequate to justify the relatively high initial expenditure in this method of providing improved pasture. It was apparent, however, during the early years of the study, that the performance of ewes and lambs on the improved pasture, particularly among twin-suckling ewes, was unsatisfactory. Development of the unit was postponed, and since 1974 it has been used to investigate the condition of hypocuprosis which was associated with the land improvement. These investigations are described in Chapter 4.

Midhill (Lephinmore)

In 1968 a study was initiated in a very different environment at Lephinmore in the west of Scotland. The Midhill unit (444 ha), rising from sea level to around 500 m, has an average rainfall of 2000 mm. Most of the hill is covered by blanket peat and the vegetation is dominated by heather (*Calluna vulgaris*), draw moss (*Eriophorum vaginatum*), and deer grass (*Trichophorum caespitosum*) with a little *Molinia*, but it is broken in places by flush areas carrying a *Juncus*-dominated grass community.

Prior to 1956 Midhill carried up to 230 Scottish Blackface ewes with weaning percentages around 55. Some enclosures were created and land improvement carried out between 1957 and 1968 and ewe numbers were increased from 205 to 372. Weaning percentage declined with increasing flock size from a peak of 98 in 1960 to 70 (Nicholson, Currie, Paterson and McCreath, 1968). In the meantime the arguments for a new system of management had been developed and Midhill provided a convenient unit on which to see whether in this environment the proposed new system would promote better performance at the increased stocking rates.

Phase 1: 1968/69-1971/72. At the start of this phase the new grazing management system was instituted and ewe numbers were reduced to 339. The improved land consisted of two enclosures totalling 69 ha; 20% of this area had been improved as a mosaic of 1-2 ha patches distributed throughout the whole (Fig. 15). The mosaic-containing enclosures were used for single-suckling ewes during lactation and for ewes during the pre-mating and

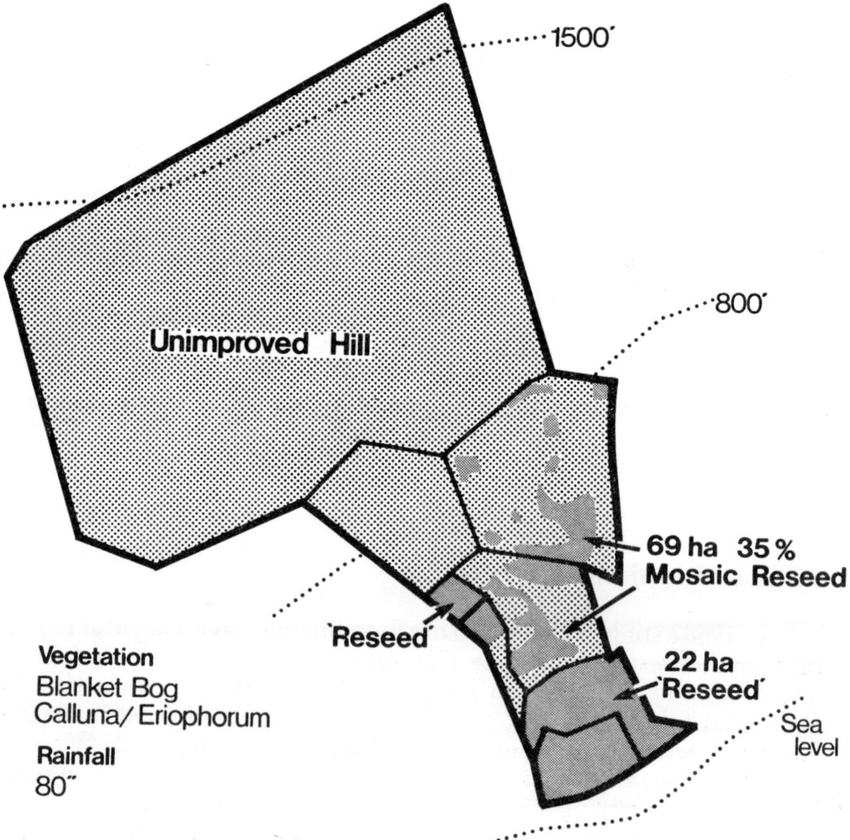

Fig. 15   Lephinmore Midhill project (431 ha).

mating period. An additional 14 ha of enclosed, permanent grassland was used for twin-suckling ewes and suckling gimmers during lactation, and again for ewes and gimmers during the pre-mating and mating period. A further 7 ha of improved land was used as a lambing paddock as well as for grazing in mid- and late lactation and during mating.

Between 1968/69 and 1971/72 ewe numbers were progressively increased to 384. Pre-mating live weights increased (Fig. 16) as did weaning percentages. The flock production data are given in Table 8.

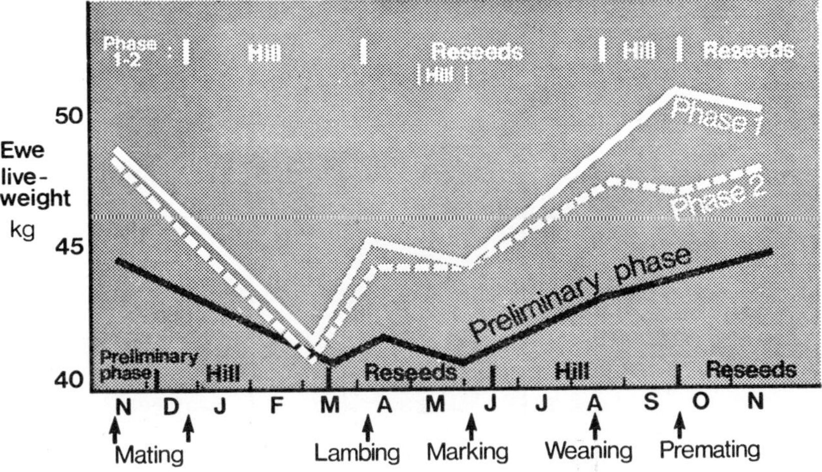

Fig. 16 Midhill, Lephinmore — mean ewe live weights for each phase over an annual cycle of production.

Ewe live weights were maintained (Fig. 16) at a much higher level throughout this phase compared to that previously (1956/57). Ewe live-weight recovery during lactation was markedly greater. Both total output of lamb and lamb output per ewe increased during this phase. Fig. 17 illustrates the relationships amongst stock numbers, individual ewe output and total flock output. It is clear that the changes in management initiated during 1968/69 had a considerable beneficial effect on both total output and individual ewe performance (Eadie, Maxwell, Kerr and Currie, 1972).

By 1971/72 ewe numbers had been increased beyond the highest level achieved in the previous study, and it was clear that individual ewe performance was very much greater than that previously achieved. The ewe live-weight data indicated that in 1972 live-weight recovery between marking and weaning was less than half that achieved in previous years and lamb weaning weights were also less. This suggested that it was an appropriate time to provide additional improved pasture.

119

Table 8:   Flock production data (Midhill)

	Phase 1				Phase 2		
	68/69	69/70	70/71	71/72	72/73	73/74	74/75
Stock numbers (ewes and gimmers)	339	361	373	384	422	433	434
Weaning percentage	85.0	92.5	103.5	103.6	103.3	98.2	91.0
Weight of lamb weaned							
total (kg)	7207	8500	10,268	9924	10,218	10,870	9638
per ewe (kg)	21.3	23.5	27.5	25.8	24.2	25.1	22.2
Total weight of wool (kg)	652	772	772	814	815	856	934

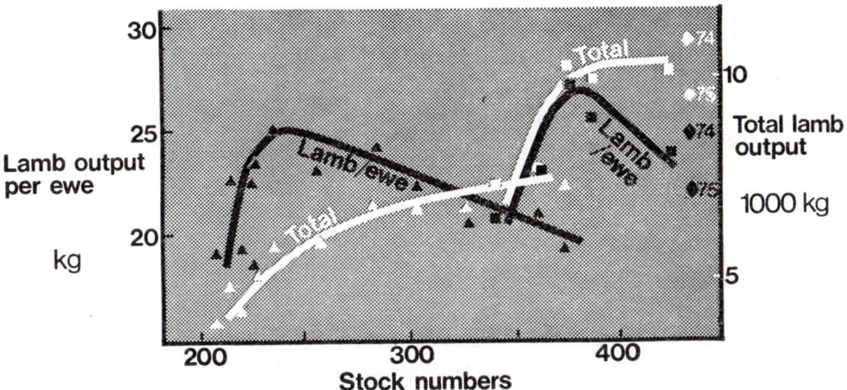

Fig. 17   Midhill, Lephinmore — relationships amongst stocking rate, total weight of lamb weaned and weight of lambs weaned per ewe mated.

Phase 2: 1972/73-1974/75. An area of 4 ha of previously improved land was incorporated into the unit in the autumn of 1972, and during 1973 a further 10 ha of indigenous pasture within the already partially improved enclosures was improved by oversowing.

During this phase ewe numbers were increased from 384 to 434. However, pre-mating weights fell from 49.9 kg to 47.9 kg with a consequent decline in weaning percentage (Eadie, Armstrong, Maxwell and Currie, 1979). Although live-weight recovery started earlier than in the previous phase, there was a reduction in the rate of recovery between marking and weaning when the lactating ewes were on improved pasture. Between weaning and pre-mating, when ewes were grazing the indigenous hill pastures they lost weight, whereas in the earlier phase they had gained weight.

Even with the land improvements in 1972 and 1973 the changes in the live-weight cycle strongly suggested that nutrition was inadequate in late lactation and that the stocking rate was approaching a level on the indigenous hill grazings which was affecting live-weight recovery between weaning and pre-mating. It was also found necessary to respond to increasing live-weight losses between mid-December and early March (when ewes are on the indigenous hill grazings) by the earlier introduction of supplementary concentrates.

Current position. Since 1975 the central objective of the study has been to establish whether the limitations to increasing output from the unit are due either to inadequate nutrition from improved pasture between marking and weaning or to inadequate recovery between weaning and pre-mating, while ewes are grazing indigenous hill pasture.

From 1968 flock gross margin had increased from £702 to £2741 and all the investments since that date have been made from cash flow. Because the original investments in land improvement were made before this system was initiated it has not been possible to measure the impact of the management strategy used since 1968 on cash flows, which in part are derived from earlier land improvements.

Barnacarry/Feorline (Lephinmore)

Barnacarry was an area of Lephinmore on which land improvement was not possible because of difficulty of access. The adjoining property of Feorline, belonging to the Forestry Commission, included a substantial area of unplantable land which was suitable for sheep grazing. The whole area of Feorline and Barnacarry was examined to see whether a re-allocation of land could be made to provide an area of plantable land, at least as large as would otherwise have been available, and at the same time provide an agricultural unit with development potential and suitable access. Subsequently, in 1974, HFRO acquired 156 ha of Feorline (71 ha deemed unplantable) and the Forestry Commission acquired 95 ha of Barnacarry (only 3 ha of which were unplantable). On the basis of this land exchange the Forestry Commission agreed to provide access roads in advance of the date they would normally be required for timber extraction. The new Barnacarry/Feorline unit extends to some 350 ha of which approximately 30-40 ha are accessible and improvable. Sheep numbers have been increased gradually from the original 227 Scottish Blackface ewes and sheep performance measured. Levels of individual performance have been low, with weaning percentages around 75, as expected on land of this quality. In response to a need to examine the performance of suckler cows in this environment and to assess sheep performance in the presence of cattle, a herd of 15 Luing cows has been established on the unit. Further development of the unit including land improvement will be considered in response to the performance of both sheep and cattle.

Cairn (Glensaugh)

The most recently initiated study is concerned with increasing the output of systems of hill sheep production on heather-dominant moorland on the Cairn unit of Glensaugh. The land resources comprise 182 ha of heather-dominant moorland rising from 200-460 m and include 23 ha of permanent grassland. There are two unimproved enclosures on the hill area used as lambing paddocks. The unit was established in 1972 and carried

234 Scottish Blackface ewes and has been managed according to the strategy outlined previously (Table 5) with hoggs being in-wintered.

The initial objective was to establish levels of production and performance to form a data base from which changes during the development phase of the study could be measured. Total lamb output has been in the region of 5000 kg with weaning percentages of about 95 and lamb weaning weights of around 25 kg. It was necessary to begin supplementary feeding in early January to prevent unacceptable ewe live-weight loss in mid-pregnancy.

One approach to increasing output from heather-dominant areas is to devise a means of increasing the utilisation of heather. Though it is possible to utilise up to 40% of the current season's shoots without affecting plant productivity, heather provides, at best, little more than a maintenance level of nutrition when it is the sole component of the diet in early summer and in winter only about half maintenance (Chapter 5). It is a significant feature of systems of hill sheep production on heather moorland, in which there is a component of improved pasture, that any improvement in the annual ewe live-weight cycle, obtained from the utilisation of improved pasture, may be lost when the ewe is grazing the heather.

Under the present system of management any increase in ewe numbers would reduce individual performance; a further input of reseeded land may improve individual performance but without necessarily improving the utilisation of heather moorland. The available evidence suggests that an association of high quality herbage in a mosaic with heather could increase utilisation of the heather up to 40% without incurring an unacceptable nutritional penalty (Milne and Bagley, 1976).

The economic implications of land improvement in this environment require an increase in individual ewe performance to justify both the capital invested in land improvement and the use of winter feeding. The continued improvement in live-weight recovery throughout the summer and autumn, which is necessary to promote an increase in output per ewe by improving conception rates, is unlikely to be sustained if ewes predominantly graze heather moorland between weaning and pre-mating. The conclusion is that some improved land must be made available during the whole summer and autumn including the period between weaning and pre-mating — a rather different approach to that already adopted.

A 20 ha enclosure of heather moorland in which 1 ha blocks of reseeded grass/clover pasture, distributed evenly throughout the whole, was established during the summer of 1978. It is proposed to make limited use of this area during lactation, with the aim of utilising only the grass areas efficiently; later it will be used between weaning and pre-mating and again, lightly stocked, during mating when the aim will be to use up to 40% of the current season's shoots of heather. A further enclosure of heather moorland containing 1-2 ha reseeded blocks will be created during the summer of 1979. The areas of permanent grass will continue to be used as previously.

Off-wintering systems programme

Two studies of off-wintering began in 1970. Within each of these, off-wintering on the one hand and off-wintering with land improvement on the other, have been compared. This was done by setting them up side by side on similar sized areas of land with similar soils and vegetation. Stock increases proceeded at the same rates on both areas. The improved pasture was integrated with the open hill in the same way as in the year-round grazing system.

Rigg and Gairs (Sourhope)

The Rigg and the Gairs are two similar units each of 101 ha and each traditionally stocked with 130-140 South Country Cheviot ewes and gimmers. The study began in 1970. On the Gairs unit an area of 15 ha of *Agrostis-Festuca* pasture was enclosed, limed and slagged early in the winter of 1969/70 and oversown with clover in the summer of 1971. In the spring of 1971, 10 ha of *Molinia/Nardus* grass heath at 450 m was limed and slagged and, in mid-July, sprayed with paraquat, rotovated and reseeded with the application of a high phosphate compound fertiliser. This was grazed for the first time in the autumn of 1971. No land improvement was carried out on the Rigg.

The flock performance data are given in Table 9.

Table 9:   Flock production data (Rigg and Gairs)

	1970	1971	1972	1973	1974	1975
Rigg						
Stock numbers	205	205	238	278	279	311
Weaning percentage	83.0	87.0	100.8	87.8	91.0	89.6
Total weight lamb weaned (kg)	3706	4432	5712	4324	6155	6257
Total weight wool (kg)	402	534	641	732	680	670
Gairs						
Stock numbers	209	207	233	260	279	305
Weaning percentage	83.0	96.0	91.4	93.1	87.0	87.2
Total weight lamb weaned (kg)	3581	5246	5176	5675	6394	6381
Total weight wool (kg)	461	524	634	752	766	732

p. 112 Figure 11

Key to treatments should read:

Key to treatments: ——————— Complete, i.e. 5t lime/ha, 3 dressings N at 40 kg/ha, trace elements, Rhizobium and 60 kg K/ha; ━━━━━━complete plus 120 kg/ha extra N (4 dressings, 40, 80, 80 and 40 kg N/ha); ••••••••••••• complete except no N added; ──────── untreated.

p. 155 Table 26

Daily live-weight gain at pasture on low herbage mass with medium level of winter feeding should read 1.02, not 1.20.

p. 170 Table 28

"Improved" should read "Unimproved".

3rd column mean should read 2.8, not 3.8.

By 1973 ewe numbers had been doubled on both units and individual performance peaked in 1972 on the Rigg with a weaning percentage of 100, and in 1971, on the Gairs with a weaning percentage of 96. At peak performance gross margins had increased to give positive net annual cash flows but were insufficient to cover the capital cost of an inwintering shed or the land improvement on the Gairs. The results achieved on the Rigg, however, suggest that an economic break-even point could be attained if it were possible to off-winter ewes and maintain similar levels of performance at these stocking rates. Further increases in stock were necessary to cover the cost of housing and land improvement. However, when stocking rates were further increased, individual performance declined and total output levelled off on both units, although, on the Gairs, pre-mating live weights had increased to 54 kg in 1974. The relationship between pre-mating live weight and lambing percentage for both the Rigg and the Gairs indicated that lambing percentage reached a peak (110) with a pre-mating live weight of 52 kg. However, since at higher pre-mating live weights with concomitant higher lambing percentages there was a higher lamb mortality, little response was apparent at weaning.

Work at the Animal Breeding Research Organisation had shown higher weaning percentages from the South Country Cheviot than were found on the Rigg and Gairs, but this was in circumstances in which nutrition favoured juvenile growth and development with a consequent increase in size and condition of the mature ewe. Levels of nutrition of these ewes during mating were higher than those likely to be attained from the pastoral resources available on the Rigg and Gairs. In addition it was found that responses in weaning percentage to increasing pre-mating weight and condition were less in the South Country Cheviot than could be expected in the Scottish Blackface (Chapter 4). In 1975 it was therefore decided to change to the Scottish Blackface breed. The complete change will be achieved by September 1979 when ewe numbers will be reduced to around 245, the number at which peak performance was achieved with the South Country Cheviot.

As a result of these measures it is already apparent that the output of lamb has increased substantially but until there is a regular-aged stock which has been bred on the units (1981) it will not be possible to make any firm conclusions.

Until 1977, 21 cattle grazed in such a way as to equate the number of grazing days per month spent on the Gairs and on the Rigg. They were maintained on these areas from 1 May to 31 December each year. By 1977 it was no longer possible to graze cattle because of inadequate pasture.

Low End (Lephinmore)

This study was conducted on a *Calluna/Eriophorum/Trichophorum*-dominant area rising from near sea level to 450 m and with an annual rainfall of 2000 mm. It had previously carried 180 Scottish Blackface ewes which produced a weaning percentage of 80 with an average lamb weight of 27 kg. Food inputs had been negligible.

In 1969 the land was divided and fenced into two areas of 167 ha (inwintered) and 157 ha (inwintered and improved pasture). The ewes were allocated equally to the two areas by weight and age. An additional 14 ha of improved 'common land' was made available to both flocks. On the inwintered and improved pasture area, 8 ha (1969/70) and 7 ha (1972) of land received lime and slag and were subsequently oversown with grass and clover seed. Compound fertiliser was applied at the time of sowing. No land improvements were carried out on the inwintered alone area.

In each year both flocks were housed from early January until lambing. For the first four years the ewes were lambed on the common area. For the remaining three years of the study the ewes were lambed inside. Both flocks were fed similarly while housed. From January until six weeks before lambing all ewes were given a basal ration of 900 g hay and 115 g cereal/protein supplement. The levels of concentrates given during the last six weeks of pregnancy were adjusted to maintain plasma ketone concentrations below 3 mg/100 ml (Chapter 4). The amounts of food given to both flocks were the same indicating that, in all years, the two flocks were undernourished to the same extent. Total food inputs during the inwintering and lambing periods averaged 80 kg hay and 21 kg concentrates per ewe.

The hoggs from both flocks were housed from the beginning of November until prior to lambing when they were returned to their respective hills. They were given some supplementary feeding for two to three weeks. They remained on the hill throughout the summer.

From 1970 until 1976 when the study was terminated, ewe numbers had been increased by 90% and the output per unit area had more than doubled. From 1971 there was a decline in pre-mating weights in both flocks though the decline was rather greater in the control flock. There were no substantial differences between the ewe weight data of the two flocks and the annual patterns of weight change were very similar. The importance of the effect of pre-mating weight on lambing performance was apparent and accounted for some 70% of the variation in lambing percentage. Growth rates of single lambs on both units declined by 10-12% during the period of the study. Twin lamb growth rates were maintained and there was little difference between flocks.

From an economic point of view the results were clear. Despite more than doubling output, the increase in costs of purchased feeding was greater than the income realised from the increased output. Even if animal performance could have been improved by a further input of reseeds (and this would have to achieve minimum pre-mating weights of 50 kg to give 110% lambs weaned), lamb prices would have had to be substantially more for off-wintering to have been economically viable in this environment. It was considered that these criteria could not be readily achieved in the foreseeable future and the study was terminated. Nevertheless, one important finding was that much higher levels of output could be obtained from hill resources of this type and that these would be economic if some cheaper means of wintering ewes could be found.

General considerations

Some general conclusions can be drawn from the hill sheep systems research programme which relate to the further development of these systems and to the Organisation's research programme. The responses in output from the longer established year-round grazing studies at Sourhope and Lephinmore provide good evidence that the concepts on which they are based have general application to the improvement of output from hill sheep farms in widely different environments.

Ewe performance

The use of improved pasture during lactation and in the pre-mating and mating period has resulted in improvement in the annual live-weight cycle, and in particular has led to increases in pre-mating weight. The studies indicate that pre-mating live weight has a significant effect on animal performance, particularly through an influence on conception rate and ultimately on weaning percentage. Where ewe numbers are being increased a point will eventually be reached where further increases depress levels of nutrition and consequently pre-mating weight and reproductive performance. The effect of nutritional limitations during the post-mating period on embryonic loss are being investigated in more detail (Chapter 4).

It is also evident that as stocking rates increase beyond a certain point ewes and gimmers tend to lose increasing amount of weight in mid-pregnancy. In the early stages of the work at Sourhope, however, it was noted that the birth weights of gimmers' lambs were poor. Results from subsequent closely controlled experiments indicated that late pregnancy nutrition was not inadequate, but that nutrition during mid-pregnancy was implicated (Russel, unpublished). These studies have been extended to examine the effects of mid-pregnancy nutrition on production from ewes and gimmers, and the possible interaction with levels of summer nutrition (Chapter 4).

Improved ewe nutrition, particularly in pregnancy and lactation, has reduced ewe and lamb mortality. Part of the increase in lambing and weaning percentages derives from the benefits of closer supervision at lambing when prompt assistance at difficult births can be provided and weak lambs can be given attention. It is, nevertheless, apparent that as year-round grazing systems become progressively more intensive they also become increasingly dependent on inputs of winter feeding to maintain satisfactory lamb birth weights. This provides the context for diet supplementation and associated metabolic studies in the animal research programme (Chapter 4).

The gimmer age group appears to be more sensitive to nutritional changes than are older ewes. Weight changes of gimmers provide a useful indication of the general level of nutrition and a useful means of anticipating the point at which further increases in stocking rate are likely to precipitate a reduction in the performance of the whole flock.

Land improvement and pasture utilisation

The key to improved animal production from all the development studies has been the introduction of improved pasture. It is important that the level of production and nutritional quality of these pastures should be maintained. Some of the improved areas on the Midhill unit at Lephinmore are more than 20 years old and although there has been a significant change in the botanical composition of the improved areas (Rogers, unpublished) levels of net annual dry matter production remain relatively high. In economic terms it is often difficult to justify the use of annual dressings of nitrogen where opportunity for further land improvement exists.

Pasture production in hill areas must involve the use of clover (Chapter 3). Problems with grass/clover in the systems development programme have highlighted issues for further research, for example, potassium deficiency on reclaimed blanket bog (Chapter 3). The maintenance of herbage production on improved hill pasture necessitates regular applications of lime and phosphate, but as fertiliser prices rise the need for a more precise statement of maintenance fertiliser requirements becomes increasingly important (Chapter 3).

The controlled grazing of *Agrostis-Festuca* in the Hairney Law/ Auchope study maintained the quality of the ingested herbage but did not significantly affect the botanical composition of these pastures. The use of cattle did little to eliminate the bracken which covered many of the gentler slopes within the enclosures. There was some doubt as to whether the elimination of bracken by spraying could be justified by the increased levels of herbage production and utilisation. This question was important in relation to the current recommendations for the approval of grants for bracken eradication. Investigations on this topic are reported in Chapter 3.

The upper limit of the ratio of improved to indigenous pasture to which the present integration strategy applies is of interest; it will differ with the nature of the hill land resources. The problem is exemplified on Midhill where, despite further inputs of improved pasture, there has been little further improvement in animal output.

It is important to recognise that, in applying the strategy in which improved and indigenous pastures are integrated in the year-round grazing studies and in which stocking rates are increased, there will come a point when the quantity and/or quality of the indigenous hill vegetation will limit further increases in animal output. It is therefore important to quantify the limits of the applicability of the present strategy. Beyond this point there are alternative management strategies but judgements as to which is the best can come only from adequate quantification and understanding.

The impact of cattle grazing on the improvement of indigenous pasture is a topic which merits further investigation (Chapter 5). The improvement of the *Agrostis-Festuca* area in the Hairney Law/Auchope study was greatly aided by the use of cattle. Their role in other nutritionally-poorer vegetational environments remains in doubt. The inclusion of cattle on the Barnacarry/Feorline unit at Lephinmore is therefore of considerable interest.

Economic considerations

Initially, inputs of improved pasture allow higher levels of individual performance to be achieved. This is currently the stage reached on the Hairney Law/Auchope unit. This study has demonstrated the possibility of increasing output on some types of hill land from a relatively small investment which, in time, has given rise to cash flows which permit further development by more intensive land improvement.

The economic viability of the year-round grazing studies on both Hairney Law/Auchope and Midhill, in terms of gross margin, net incomes and return on investment, has been well demonstrated (Armstrong *et al*, 1978; Eadie *et al*, 1979). Responses to investment were, however, relatively slow and cash flows in the early years were negative. Patterns of investment are very important in terms of cash flow, and the re-introduction of the systems study on Alderhope, in which a large investment in pasture improvement was made over a relatively short period, will provide some contrasting data on this aspect.

There inevitably comes a point in the intensification of year-round grazing systems when the contribution of the indigenous pastures to winter nutrition is severely reduced. The economic viability of such a situation will depend on the relative values of purchased food, store lambs, cast ewes and wool, and the extent to which ewe numbers might be further increased as a consequence of a lesser dependence on pastoral resources during a large part of the winter. The next logical development is off-wintering, with its advantages of greater control of nutrition, and the more efficient use of labour, which constitutes a high proportion of the fixed costs. The capital investment required is high; levels of output must be correspondingly high to service the capital investment over a relatively short period of time.

Return on capital is not necessarily the most important criterion by which a farmer judges his investment; he may be more interested, for example, in the levels of income generated. Much depends on whether or not he is borrowing capital and, of course, there are alternative ways in which he could invest his own capital. Even although a short-term improvement in income may be generated, it would be imprudent to accept a return on marginal capital investment which is significantly less than could be obtained from alternative forms of investment.

Health control

It was recognised at the inception of the development programme that, unless disease could be rigorously controlled, the potential increases in output might not be fully realised. Consequently, strategic preventive medicine programmes including clostridial and louping-ill vaccination, anthelmintic medication, copper adminstration, routine dipping and foot bath treatment, were instituted as appropriate for each of the studies. Faecal and blood samples were collected regularly and all outbreaks of disease were closely investigated. Investigations of the effects of some important diseases on productivity have been carried out. In setting up the programme of monitoring and disease control, account was taken of diseases known to be endemic, of current preventive practices, and of possible changes in the pattern of these and other diseases which might emerge as a consequence of intensification.

Amongst the first conditions to be considered were parasitic diseases, including those caused by nematodes and liver fluke and those transmitted by ticks. Work in this area included regular monitoring and dosing trials to test the efficacy and timing of the use of anthelmintics. Preventive programmes were adjusted as required. Nematodiriasis outbreaks, not normally associated with hill sheep farming, occurred on reseeded pastures used each year by lambs during the critical period. Monitoring of parasitic disease has included pasture larvae counts, the use of 'tracer' lambs and the setting up of plots to forecast outbreaks of nematodiriasis on each research station. While the use of improved pastures may produce problems atypical of traditional management systems, it has been possible to keep these well under control.

An outbreak of fascioliasis at an early stage of the Hairney Law/ Auchope study led to routine dosing with flukicides and the use of molluscides on known snail sites. The improved pastures at Lephinmore created a favourable environment for the intermediate snail host of the liver fluke, and increased populations of these snails were observed. Liver fluke was endemic on Lephinmore and at one time approximately 80% of all ewes were passing fluke eggs. A programme of strategic dosing with a flukicide capable of killing both mature and immature flukes, and with dosing intervals designed to prevent the development of immature stages to egg-laying adults, was introduced in 1973. By 1977 the proportion of ewes passing fluke eggs was reduced to less than 1%; no clinical cases of fascioliasis have since occurred, health has improved and liver fluke is no longer endemic or even sporadic at Lephinmore (Whitelaw and Fawcett, 1977).

The use of lime, compound fertilisers and the introduction of new species of grass into hill soils required the monitoring of the trace element status of different groups of sheep to identify the areas and seasons of risk. This led to the confirmation of suspected hypocuprosis on Alderhope and ultimately to a detailed study of the problem reported in Chapter 4.

The programme of monitoring continues and veterinary investigation reports are used to identify potential risks. This information will continue to be used to develop preventive medicine programmes which are tailored to the requirements of individual flocks. Flexibility is maintained; sometimes preventive measures require to be increased and at other times progress in control can lead to less intensive measures although monitoring continues as a safeguard.

Mathematical modelling

Mathematical modelling is seen as a means of extending the scope of systems research and providing a dynamic, quantitative description of the biological components of production. While it is essential to carry out analytical research to understand how the separate biological components of animal production function, the understanding is limited by the way in which analytical research is necessarily conducted. Systems performance cannot be judged simply in terms of how each part works separately but must take account of how the parts relate to each other and of how the system relates to its environment and to other systems. The implication is that research should be conducted not merely on the basis of analysis but also on the basis of synthesis. Mathematical modelling provides a means whereby a synthesis may be developed. Although field testing will always be required, the results from mathematical modelling may reduce the number of options that require to be tested in this way.

Systems can be modelled at various levels of biological organisation and also in economic terms. The level chosen reflects the size and purpose of the system, for example the cell, organ or whole organism. For the purpose of modelling, a system can be described in terms of a commodity (for example, energy) which can exist in a number of discrete states (for example, energy in pasture or in the animal) and for which rates of transfer can be derived (for example, the rate of intake of pasture by the animal). Management systems are concerned with controlling these rates of transfer.

The synthesis of agricultural systems is essentially concerned with arranging the use of resources to achieve a particular objective. The objective is often to increase the efficiency of production, i.e. to reduce the level of inputs for each unit of output. The units may be expressed as money, energy, protein etc. In a well defined and understood system it will be possible to define the means of achieving this objective in precise quantitative terms.

The broad strategy of the modelling programme in the Organisation was conceived as a means of obtaining a clearer and less ambiguous understanding of biological processes and mechanisms and, where possible, to provide a quantitative description of them. These models were also used to build larger models of animal production systems in order to examine the effect of varying the levels of inputs, alternative management procedures and their economic viability.

Some models are principally concerned with specific biological components of production. Previous reference has been made to the development of a grazing model (Chapter 5), a model of lactation is described later (Chapter 7) and a pregnancy model is being developed. Another model to provide a more objective basis for the regulation of supplementary feeding in late pregnancy, in response to changing concentrations of certain blood metabolites, is being tested (Chapter 7). The process of building models has contributed towards the formulation of hypotheses and identification of gaps in knowledge.

Land use model

There is much debate on the subject of land use in the hills and uplands, particularly in relation to agriculture and forestry (Cunningham, Eadie, Maxwell and Sibbald, 1978). Present policies require that all land to be acquired by the Forestry Commission for afforestation and proposed private plantings under the Dedication Schemes should be investigated by the relevant Government departments. Any disagreement between the Forestry Commission and the departments over the relative advantages of taking land for afforestation or leaving it in agriculture is submitted for decision at Ministerial level. The guidelines on which recommendations are based leave much to the judgement of the Land Inspectorate (GB Parl, 1972). This has led to dissatisfaction amongst both agriculturalists and foresters concerning decisions which they see as inhibiting their operations.

A modelling approach to the problems of land allocation has the advantage that rules and priorities are explicitly defined. Their effect and significance can be examined in terms of several objective functions, for example national economic return, return to agriculture (the farmer), the return to the forester (Forestry Commission or private forestry) and labour requirements. The model is still under development (Maxwell, Sibbald and Eadie, 1979) and a number of case studies are being carried out in co-operation with the Department of Agriculture and Fisheries for Scotland and the Forestry Commission. The existing model has been designed for use on a specific parcel of land representing an agricultural holding of a size capable of sustaining both agricultural and forestry interests. It is hoped to extend the model so that several adjoining holdings may be examined.

In the development of the model it has been accepted as fundamental that the land allocated to agriculture allows the creation of an economically viable unit. It has also been accepted that the entire agricultural unit has to be contiguous. Land that is close to existing farm buildings may be preferentially selected for agriculture. The model minimises the amount of fencing and road building by the appropriate aggregation of blocks selected for agriculture where possible.

Use of the model has demonstrated that in some circumstances the economic return from land which is used by both agriculture and forestry can be greater than when only one of the industries uses that land. It has also demonstrated that very different land allocations, for example, where agriculture was placed on the best land in one case and the poorest in the other, could give rise to similar results in terms of the total return from both industries. The choice of discount rate will also influence the results obtained.

Economic model

An economic evaluation of alternative systems is clearly important in the development of hill and upland agriculture. Econometric models can be described as comprehensive calculators which represent the inputs and outputs of a system in financial terms.

In the development of hill sheep production systems in the Organisation, capital investment was required and changes in variable and fixed costs were likely to follow. An economic model was developed which allowed an examination of the effect of varying patterns of capital investment, changes in variable and fixed costs, and changes in stock numbers on gross margin, net annual cash flow and return on investment (Maxwell *et al*, 1973a; Sibbald and Maxwell, unpublished). The changes in animal performance and output necessary to justify the capital investments and changes in variable costs could then be ascertained. Alternatively, the amount and timing of capital investment, possible on the basis of specific levels of animal performance and output, could be calculated.

The feasibility of any particular scheme remained a matter of judgement until data from the systems programme at HFRO and from other centres became available. Sensitivity analysis, using varying price/cost relationships, provided a means of testing the robustness of particular schemes. With increasing data and experience of the response in animal output to capital investment, economic assessments can now be made with a greater degree of confidence.

Some hypothetical studies made in the early seventies suggested that comparatively small amounts of capital, invested in land improvement for example, could give rise to quite substantial cash flows and involve only short periods of negative balance. Large amounts of capital on the other hand, particularly if invested all at once, could produce very difficult cash flow problems. High levels of capital investment clearly need to be spread over a number of years and made at times closely related to the expected increases in animal performance.

These conclusions suggested a strategy for improving net farm incomes by means of comparatively low levels of capital investment. The investment, after a few years, was likely to give rise to increasing year-end balances and further investment could be made from cash flow. Even if the return on this capital was low (for example, if the first step was to in-winter or off-winter) it could give rise, ultimately, to a large increase in farm income and in business size. Alternative policies could be compared on the basis of net present value to determine which makes the best use of capital. At all stages in the development of a project these methods have proved invaluable and interest in their wider application to commercial farm assessment has recently become apparent.

References

ALLEN, G R (1973). The economic outlook for hill farming in the 1970's. In: *Hill pasture improvement and its economic utilisation.* Colloquium Proceedings No 3, Potassium Institute Ltd. Edinburgh, 1972, 55-62.

DUTHIE, W B (1968). *Financial results, East of Scotland farms 1966-67*. Edinburgh School
of Agriculture Bulletin 90.

GB PARL (1972) HC 511-v. Select Committee on Scottish Affairs Session 1971-72. *Land
resource use in Scotland*. Vol. 5 Appendices to the minutes of evidence and index.
London, HMSO, 11-13.

HARKINS, J (1968). Assessing new capital investment on hill sheep farms. *Scott Agric 47,*
196-200.

HARKINS, J (1970). Amalgamate or intensify? *Blackface Sheepbreeders Assoc J 22,* 26-29.

LORD, R F and BLYTH, A (1973). Trends in profitability and resource use on hill farms in
SE Scotland. In: *Hill pasture improvement and its economic utilisation*. Colloquium
Proceedings No 3, Potassium Institute Ltd. Edinburgh, 1972, 87-94.

7

Upland sheep research

Contributors: T J Maxwell, A J F Russel, A R Sibbald
Co-ordinators: R G Gunn, J Eadie

Introduction

Successful commercial managers of pasture-based systems of animal production are those who have learnt by experience how to balance the needs of their stock with the need to ensure an efficient use of the pasture. In the course of the sheep year many management decisions have to be made, for example when to move sheep from one pasture to another; when and how much fertiliser to apply; when to introduce supplementation and how much to feed; when to withdraw supplements; how much land to enclose for fodder conservation; and so on. In any commercial system it is difficult enough for the farmer to make the best decisions in relation to these factors, even within the context of a previously chosen stocking rate and lambing date. Introduction of further changes in the latter factor increases the complexity of the decision-making process.

The aim of research must be to put the necessary decision making on a secure, objective and predictable basis. It is helpful to distinguish between the issues which can be resolved more appropriately by means of analytical or component research and those which should be considered as systems questions. Into the latter category come those major decisions, such as choice of stocking rate and lambing date, whose consequences clearly affect the system in a continuing way throughout the year. Other factors such as pregnancy nutrition, although they also have consequences for the whole system, are initially best studied by the analytical approach.

The Organisation's approach to research on upland sheep is therefore based on both systems studies and component research, and a deliberate attempt is made to interrelate the two kinds of activity. This has been achieved by carrying out experiments on specific aspects of sheep performance within the framework of a systems study and by incorporating the information obtained from the component research programme back into the development of systems. The priorities are determined in relation to both programmes. In the following outline of progress to date, the framework of the discussion is derived from the systems research programme. Although the analytical research which has proceeded in parallel with this work is discussed within this framework it also makes a contribution to the general body of knowledge of sheep biology and production.

Systems experimentation

It is clearly necessary to examine the relationships among the various components of production within year-round systems and to manage these systems using rigorously defined decision rules. By this means, the effects of decisions made at one time, and which have consequences not only at that time but throughout the annual cycle of production, can be seen within the perspective of a whole system. Interpretation of results in relation to the separate components of production becomes possible and a context is provided in which these components can be examined in depth. Because the system is investigated as a whole and not just in terms of its component parts, the important interactions amongst these components can be identified, for example between pregnancy, lactation and reproduction. If systems experimentation is conducted over a sufficient number of years it is also possible to quantify the effects of a system on the lifetime performance of the ewe. It was with these issues in mind that a programme of research to study systems of upland sheep production was initiated in 1972.

From a review of existing knowledge it was clear that output from sheep utilising sown grassland is influenced by the seasonal pattern of pasture production, by stocking rate and also by their interaction. The choice of lambing date at any stocking rate will affect the relationship between nutrient provision and nutrient need and will therefore have production consequences. The choice of stocking rate and lambing date will also have an influence on the amount of surplus pasture that can be conserved to provide winter feeding and on the amount of pasture available to ewes and lambs during the period of closure. Stocking rate and lambing date were therefore the first major issues examined in the systems experiment which was conducted from 1973 to 1977.

A study of four systems was established at Glensaugh where an area of 19 ha of permanent or semi-permanent pasture at 230 m was used for spring, summer and autumn grazing. This area comprised four fields sown out in 1954, 1967, 1972 and 1973. A further 14 ha of hill land dominated by heather (*Calluna vulgaris*), but interspersed with *Agrostis-Festuca*, were grazed from January until lambing. Two stocking rates (approximately 10 and 15 ewes per ha) and two times of lambing (early March and early April) were examined. Each field was divided into four similar paddocks and one in each field was allocated to a stocking rate/lambing date treatment. The input of nitrogen was limited to 125 kg/ha for all systems. The study was conducted with 225 Border Leicester × Scottish Blackface (Greyface) ewes mated to Dorset Down rams.

Derivation of decision rules

A major problem which had to be resolved at the outset was to devise a procedure which would permit adequate comparison of the results of each of the stocking rate/lambing date combinations. This was done by a decision rule procedure. Each decision rule is founded on some aspect of the intrinsic biology of the system and therefore reflects in part the consequences of an individual 'treatment'. When taken together, the decision rules describe the system and allow valid comparisons to be made. It is not claimed that the decision rules are necessarily the best that can be devised, nor that in all cases sufficient information exists on which to found an objectively-based rule. Some require a compromise between total objectivity and practicality in that a manageable programme of measurement has to be carried out. What can reasonably be claimed, however, is that the need for such decision rules has been recognised and that a vigorous attempt has been made to provide them.

The decision rules are derived from studies relating the component research and the systems experimentation and are described later following presentation of results.

Results

A summary of the production data from each of the systems is given in Table 10.

Table 10.   Production data 1974-77

Lambing date	Stocking rate	Lambs weaned (%)	Lamb weight (kg/ewe)	Lamb weight (kg/ha)	Wool (kg/ewe)
Early	low	148	44.6	420	2.49
Early	high	150	42.9	646	2.69
Late	low	160	51.7	488	2.58
Late	high	152	45.7	681	2.46

There was considerable difference in the weight of lamb weaned per unit area from the low and high stocking rates. The total lamb output of the later-lambing groups was greater than that of the early-lambing groups as a result of differences in lamb weaned per ewe. These differences arose mainly through lamb mortality rather than reproductive performance, although growth rate and weaning weight also contributed.

Table 11. Reproductive performance 1974-77

Lambing date	Stocking rate	Pre-mating weight (kg)	Mating weight change (kg/28d)	Lambs born (per 100 ewes mated)	Lamb deaths (%)	Lambs weaned (per 100 ewes mated)
Early	low	71.3	+1.14	178	14.2	148
Early	high	68.7	+0.38	176	13.7	150
Late	low	68.1	−1.28	171	6.2	160
Late	high	66.5	−2.23	165	8.4	152

Reproductive performance is shown in Table 11 in relation to pre-mating weight and weight change during mating. The results suggest that the effect of weight change during the period of mating is important in determining the number of lambs born.

Although early-lambing treatments resulted in more lambs born they also had a higher level of lamb mortality. The late-lambing treatments therefore weaned the greater number of lambs per ewe mated.

Table 12. Lamb birth weights (kg) 1974-77

Lambing date	Stocking rate	Singles	Twins
Early	low	4.80	4.21
Early	high	4.99	4.22
Late	low	5.20	4.26
Late	high	5.32	4.32

There were no real differences between treatments in the birth weights of twin lambs (Table 12) but single lambs, which were relatively few in number, were heavier in the late group than in the early group.

Lamb weaning weights at 16 weeks are shown in Table 13. The only real difference between lambing-date treatments was an advantage of about 1 kg in the case of twin lambs in the late group but differences between

Table 13.   Lamb weaning weights (kg) 1974-77

Lambing date	Stocking rate	Singles	Twins
Early	low	36.13	30.38
Early	high	34.31	28.59
Late	low	37.61	31.34
Late	high	33.98	29.75

stocking rates were greater for both singles (2.74 kg) and twins (1.70 kg). These differences contributed to the greater output of lamb per ewe from the low stocking rate treatments and the late-lambing groups.

Table 14 provides details of the food inputs to each treatment and the amount of hay conserved within each.   Pre-partum concentrate feeding was very similar for both early- and late-lambing treatments and did not differ between the stocking rate groups within these treatments. The major difference in feed inputs occurred during the post-partum period when the early groups required some 40 kg more concentrate than the late groups in early lactation. This difference demonstrates the importance of the date of initiation of pasture growth. It was always possible to conserve from within the system all the hay required for both low stocking rate treatments. In the high stocking rate treatments it was possible to make part of the hay required for the late-lambing group, but no hay was ever made in the early-lambing treatment.

Component research

This section discusses the research carried out into various aspects of upland sheep performance and pasture use to provide information required for the formulation of the decision rules.

Pregnancy nutrition

It was necessary to find an objective method whereby inputs of winter feeding could be adjusted to take account of between-year differences in food quality, variation amongst treatments in weight and condition, and in the number of lambs conceived. Previous work on hill ewes had demonstrated that it was possible to regulate the quantities of supplementary feeding to achieve the acceptable nutritional state required to ensure satisfactory lamb birth weights. This was done on the basis of circulating concentrations of blood metabolites (Chapter 4). There was, however, little information available for the breeds and crosses traditionally used in the

Upland sheep at Glensaugh

Table 14. Food inputs and total hay conserved and consumed

Lambing date	Stocking rate	Concentrate food inputs (kg/ewe)			Hay fed (kg/ewe)			Total hay fed (t)	Total hay made (t)
		Pre-partum	Post-partum	Total	Pre-partum	Post-partum	Total		
Early	low	33.5	78.3	111.8	63.2	24.9	88.1	4.39	9.00
Early	high							6.26	Nil
Late	low	34.3	38.0	72.3	80.8	10.1	90.9	4.00	8.00
Late	high							6.02	2.24

uplands. Initially, therefore, an attempt was made to ensure that lamb birth weights were not reduced by more than 10%, by using the information previously derived from hill ewes. Research was simultaneously undertaken with the more productive and heavier Greyface ewes to obtain information more relevant to upland sheep.

The results indicated that a moderate degree of undernourishment in late pregnancy, as characterised by plasma 3-hydroxybutyrate (3-OHB) concentrations in individually fed ewes of about 1.1 mmol/l, was likely to reduce lamb birth weights by less than 10%. Lamb birth weights which are some 8-10% less than those from ewes fed to meet their full energy requirements in late pregnancy are not likely to have any significant adverse effect on subsequent performance, and almost certainly not a sufficient effect to justify the additional energy intake required to prevent such a reduction. A more severe degree of undernourishment (plasma 3-OHB concentration of about 1.6 mmol/l) resulted in birth weight reductions of about 25% which would almost certainly prejudice survival and might also affect subsequent growth rate. The cost of this production penalty is likely to be greater than the price of the feeding stuff saved. Such results were very similar to those obtained with hill ewes (Chapter 4).

The above results were derived from work undertaken on individually-fed ewes and, when seeking to apply them to the flock situation, account must be taken of factors likely to contribute to between-animal variation in nutritional state. The interaction of the components of variation in nutritional state is such that if the prescribed degree of undernourishment known to give satisfactory lambing performance in individually-fed ewes is applied to the flock situation a proportion of ewes, those with larger twins and triplets, will be severely undernourished. To avoid this the only practicable action at present is to reduce the prescribed degree of undernourishment of the flock as a whole. For an acceptable compromise between uneconomically high energy inputs and excessive reductions in lamb birth weight, a nutritional state characterised by a mean plasma 3-OHB concentration of 0.8 mmol/l was considered to be appropriate to the flock situation rather than the 1.1 mmol/l applied to the individually-fed ewes. Using this value, the birth weights of lambs within the systems studies were consistently satusfactory in all years and treatments (Table 12).

The research on pregnancy nutrition in upland ewes has provided important information on two other aspects. Firstly, it provided an estimate of the energy requirement of the products of conception, with the value of 1.5 MJ ME/kg foetus/d in the final days of pregnancy being identical to that established in earlier work with Blackface ewes (Russel, Doney and Reid, 1967b). Secondly, it enabled quantitative relationships to be established between energy intake and indices of nutritional state. These relationships showed, for example, that energy balance is represented by a plasma 3-OHB concentration of 0.7 mmol/l. They also allowed the amount of additional energy required to change 3-OHB concentration from the measured to the prescribed value to be calculated in ewes of different live weights. By adding to the equation a term for the expected increment in foetal weight over a specified time interval, it is possible to determine the additional energy required to control the deficit to within defined limits over the specified period at different stages of pregnancy. The results of this work are now the basis of a computer program which calculates the changes in the quantity of supplementary feeding required on either an individual or flock basis for ewes varying in live weight and expected lambing percentage.

Further studies have been undertaken to examine the between-animal variation in intake on plasma 3-OHB concentration in group-fed animals. The possibility of a differential rate of fat mobilisation between fat and thin ewes has been considered and whether this, if demonstrated, would justify the use of differential levels of undernourishment based on live weight or condition at the start of the feeding period.

Lactation and lamb growth

Many factors influence the milk production of ewes (Chapter 4). These include breed, number of lambs suckled, body condition at parturition, diet quality and rate of expansion of voluntary intake. As with pregnancy nutrition, however, little information on lactation performance, on associated lamb growth rates, or on factors influencing these components of production was available for upland genotypes. Studies were therefore carried out on Greyface ewes to obtain the relevant information.

Nutrition. In one study, the effects of rearing type and prepartum nutrition on the voluntary intake and milk production of lactating Greyface ewes at pasture was investigated (Maxwell, Doney, Milne, Peart, Russel, Sibbald and MacDonald, 1979). During late pregnancy the ewes were fed indoors on a range of hay and concentrate diets designed to severely undernourish, moderately undernourish or meet the full energy requirements (3-OHB concentrations of 1.6, 1.1 and 0.7 mmol/l respectively). When turned out to pasture at the time of parturition both lactating and non-lactating ewes increased their organic matter intakes, the former to the greater extent. Over the first seven weeks lactating ewes consumed on average between 40% and 50% more than non-lactating ewes.

Although both single- and twin-suckling ewes which had been severely undernourished ate slightly more than ewes which had been moderately undernourished or adequately fed, this had very little effect on milk production or ewe live-weight change. Single-suckling ewes gained a little more weight than twin-suckling ewes which produced more milk.

All ewes gained weight at the start of lactation irrespective of the pre-partum level of nutrition and the gain varied in relation to the energy available after meeting the demands of milk production. It can be assumed that the amount of herbage available in this experiment was not limiting and, therefore, the results indicate the potential milk production of this cross-breed at pasture.

Although one objective of the systems experimentation was to maximise animal performance through the utilisation of sown pasture, there may be a need to introduce supplementary feeding during early lactation when herbage growth is often limiting. Initially it was assumed that ewes would progressively restrict their intake of supplement as herbage weight increased, but the results of this experiment showed that they were just as likely to substitute as to supplement their diet of grass. Before this information was available, however, an arbitrary decision rule on the amount of supplementation and the time of withdrawal was rigorously applied.

Further information was clearly needed on the effect of varying amounts of supplementary feeding and herbage on the organic matter intake of ewes at grass. An experiment was therefore carried out in which the effect of level of concentrate and weight of herbage available on the intake and performance of ewes with twin lambs at pasture in early lactation was examined. There was no response in lamb growth from supplementation at lower levels of herbage weight (500 and 750 kg DM/ha). Ewe live-weight change was related to amount of supplement fed at only the lowest level of herbage weight (500 kg DM/ha). Differences in ewe and lamb performance were related to total organic matter intake of the ewe and to differences in the digestibility of ingested herbage (Milne, Maxwell and Souter, 1979).

The growth rate of lambs during the first six weeks of lactation in the systems experiment differed markedly between the lambing-date treatments but were similar thereafter. The growth rate of twin lambs in the early-lambing treatments was less than in the late-lambing groups over the first six weeks, but the early-lambing ewes ate more than twice as much concentrates as did those lambing later. These results also demonstrate the importance of herbage weight, irrespective of concentrate use, on lamb growth and ewe live-weight recovery.

Conservation and stocking rate. The impact of removing areas of pasture for conservation on the amounts of herbage remaining for grazing during mid-lactation may be critical. In the early stages of the systems experiment and in the absence of any quantitative data on the effect on performance of herbage weight per hectare or allowance per ewe, the areas to be used for conservation were determined by the arbitrary decision to maintain a minimum herbage allowance of 45 kg DM/ewe. While lamb growth rates were generally satisfactory the amount of hay conserved using this rule gave

Concentrates

Blackface sheep
at Lephinmore

Supplementary feeding

Hay

Red Deer at Glensaugh

rise to surplus winter forage in both low stocking rate treatments, suggesting that a greater herbage allowance or higher stocking rate could have been used. Conversely, in the high stocking rate treatments the supply of winter forage was clearly inadequate, suggesting that a lesser allowance or lower stocking rate could have been used. The choice of level of herbage allowance affects not only lamb growth rate but also affects ewe live-weight recovery and conception rates. These effects must be viewed against the substantial economic benefit of conserving sufficient winter forage. Minor adjustments to the level and pattern of fertiliser inputs may enable high levels of individual performance, high levels of output per unit area and the production of adequate winter forage to be more effectively harmonised. More information is required on the effect on animal production of herbage weight per hectare and allowance per ewe to establish the most effective level for the decision rule on conservation.

Modelling lactation. The control of lamb growth rate through pasture management to influence ewe nutrition and level of milk production is complex. A better understanding and quantitative description of lactation is required before a more precisely controlled grazing management system can be formulated. An attempt to clarify some of the issues and to produce a statement of existing knowledge, using a modelling approach, has been undertaken.

It has been assumed in the model that there is a maximum potential level of milk production related to the number of lambs suckled and to the breed of ewe. During early lactation it has been assumed on the basis of earlier work that if nutrient intake, above that required for maintenance, is inadequate for potential milk production, this will be achieved only if the ewe can draw on body reserves (Chapter 4).

The model has been designed in relation to a grazing model (Chapter 5) in which energy intake derived from pasture is limited by the quality and amount of herbage and by the extent to which a lactating ewe can expand its appetite. Although the level of expansion may be related to diet quality, level of milk production or the body condition of the ewe, the available quantitative data are inadequate to model appetite expansion satisfactorily. The model merely adopts levels of expansion which have been observed in experiments.

Energy partition in early lactation, when intake may be restricted, has so far caused the greatest difficulty and it has only been possible to do this deterministically. As lactation proceeds the partitioning of energy surplus to maintenance requirement is increasingly towards live-weight recovery.

The model is therefore built on certain assumptions which themselves have not yet been quantitatively validated. The central issue appears to be the identification and quantification of the factors that control milk production. It is known that milk production is affected by all the factors detailed, but the way in which they are related and the circumstances in which any one is likely to be more important than others has yet to be established.

Reproduction

Results on reproductive performance in the systems experiment are shown in Table 11. Analytical studies on the reproductive capability of the upland ewe have been recently instituted. Many of the principles that control reproduction have been established in relation to the hill breeds (Chapter 4) but there was a lack of information on the detailed responses of breeds and crosses suitable for upland systems. The use of first crosses between hill and lowland breeds, the purchase of replacements at six or 18 months, the more intensive grazing management and the higher levels of reproductive performance expected, all pose new questions.

Evidence in the Blackface breed suggests that growth during development and size at maturity are important determinants of life-time reproductive performance. Although the crossbreds used in upland situations have a relatively high reproductive potential their early growth is generally determined outwith the system in which they will spend their productive life. Systems of upland sheep management which rear their own replacements may therefore confer advantages to the control of future performance. One such system is being tested at Glensaugh. The pure-bred maternal hill stock are being maintained and both first cross ewes and second cross lambs are being produced within one flock. This work is still in its early stages and results are not yet available.

Time of mating in upland systems is usually determined by the aim to produce an early lamb crop. Early mating may not be successful in some years as not all ewes will have commenced their breeding season. In the systems experiment a delay in the onset of the breeding season until late October was observed in two out of five years. Attempts to explain this in terms of either nutritional or stress factors have so far been unsuccessful. Nevertheless, the earlier in October that Greyface ewes can be mated the greater the potential lambing rate. This is achieved partly through a greater ovulation rate and partly through a lower rate of embryo loss associated with reduced weather stress. Earlier mating does, however, introduce problems associated with nutrition and climate in late pregnancy, at lambing and in the early life of lambs.

The effects of nutrition, both indirectly through achieved body condition and directly through the level of intake immediately prior to and during mating influence ovulation rate and eventual lambing percentage as in hill breeds. Studies on the effect on ovulation rate of body condition at mating in Greyface ewes have shown a decline from about 2.6 at condition scores 3.0-3.5 to about 1.7 at scores 1.5-2.0.

Short-term nutritional effects have also been studied indirectly in the systems experiment at Glensaugh. Reproductive performance was closely related to the direction of live-weight change about the time of mating and therefore to the nutritional circumstances at that time (Gunn and Maxwell, 1978). Ewes losing weight at mating produced fewer lambs (1.58 per ewe lambing) than ewes maintaining (1.78) and gaining (1.96) weight. The differences in reproductive performance between times of mating and between years can be explained largely in terms of the proportion of ewes in the different weight-change categories.

From other studies with Greyface ewes it appears that the direction of live-weight change before mating may be more important than that post-mating, although whether the effect is through ovulation rate or early embryonic wastage, or both, is not clear. It is possible, however, that wastage rather than ovulation rate is affected by the level of nutrition prior to mating, since evidence from studies on the hill breeds (Chapter 4) has shown ovulation rate to be influenced by pre-mating nutrition only at relatively low levels of body condition — levels far below those observed in the Greyface ewes in the Glensaugh systems experiment.

Decision rules

The decision rules derived from the foregoing studies on the relationships between component research and the systems experimentation are summarised below.

(a) **Grazing management.** The ewes in each system are set stocked. Every two to three weeks the amount of herbage available in each paddock is measured and the ewes are allocated to the four paddocks in each system to equalise the herbage allowance per ewe.

(b) **Concentrate feeding.** In late pregnancy quantities of supplementary feeding are adjusted weekly to maintain mean plasma 3-OHB concentrations at not more than 0.8 mmol/l.

In early lactation quantities of supplementary feeding are adjusted according to the amount of herbage available. At levels less than 500 kg DM/ha, 1.5 kg concentrates/ewe are fed; between 500-600 kg DM/ha, 0.5 kg concentrates/ewe are fed; above 600 kg DM/ha feeding is stopped.

In the pre-mating period and for 34 days after the beginning of mating, if the amount of herbage available is less than 500 kg DM/ha, sufficient feeding is given to ensure that the ewes' maintenance requirements are met.

(c) **Conservation.** The date on which areas of not less than half a paddock are enclosed for conservation is determined by the date when it can be assumed that the 100 mm soil temperature will not fall below 5.5°C for the rest of the growing season (this date is calculated as the day when the accumulated degrees by which the daily 100 mm soil temperature exceeds 5.5°C reach 23) providing that this occurs before 20 May and that the herbage allowance on the remaining area exceeds 45 kg DM/ewe. An area intended for conservation is reopened for grazing if the herbage allowance on the grazing area falls below 45 kg DM/ewe during the period from 20 May to 17 June. Conserved areas are harvested between 20-25 June, and thereafter opened for grazing when the herbage regrowth reaches 500 kg DM/ha.

(d) **Fertiliser application.** On pasture which is not required for conservation an initial application of 62 kg N/ha is given when the 100 mm soil temperature reaches 5.5°C. The date of the second application (31 kg N/ha) is calculated as six weeks after the mean date of lambing, plus the number of days after the first application when it can be assumed that the 100 mm soil temperature will not fall below 5.5°C for the rest of the growing season (see above). The third and final application of 31 kg N/ha is given at weaning.

On areas to be used for conservation two fertiliser applications are given — the first (93 kg N/ha + P + K) when the 100 mm soil temperature reaches 5.5°C, and the second (31 kg N/ha) after conservation.

8

Beef cattle research

Contributors: J Hodgson, A J F Russel
Co-ordinators: J N Peart, J Eadie

Introduction

In 1972 the Organisation's remit was extended to include the problems of upland farming. The initiation of a programme of research on the suckler beef cow and calf was seen, therefore, as a logical development.

The overall objective of this research programme is to quantify and understand the biological factors influencing productivity to improve the efficiency with which hill and upland resources are used for suckler beef production. Two closely related lines of research have developed, one nutritional and the other concerned with aspects of grazing and pasture utilisation.

As regards nutrition, the largest single item of cost of producing a weaned beef calf is that of feeding the suckler cow during the winter months. This is especially important in hill herds where the problem of providing winter keep is often particularly acute. Ideally, the winter food inputs to suckler cows, whether pregnant or lactating, or both, should be the minimum which will ensure a satisfactory level of production in the current year and which will not adversely affect production in subsequent years. To obtain a clear understanding of what constitutes such acceptable levels of feeding for cows at different stages of pregnancy and lactation, it is necessary to study and quantify the relationships between nutritional inputs and the components of production with cows in different physio-logical states.

The interest in grazing studies arises from a need to understand the interactions between the animals and the pastures, hill or upland, which constitute the basic nutritional resource. Some of this work, for example the evaluation of indigenous and sown pastures as sources of nutrients for beef cows and the study of factors which influence diet selection and intake, is described in Chapter 5.

The research programme

Phase 1: 1973-75

The first phase of the programme began in 1973 when the autumn-calving herd at Glensaugh was built up to around 120 with approximately equal numbers of Hereford × Friesian and Blue Grey (Whitebred Shorthorn × Galloway) cows. Hereford bulls were used in 1973 and 1974, but since then the cows have been bred to Charolais bulls. A new cattle house with 128 individual stalls was built.

In the first two years the work was concerned with examining the effect of nutrition in late pregnancy on calf birth weight, body condition and subsequent performance. A specially prepared roughage diet was fed at levels ranging from 65% to 170% of non-pregnant maintenance requirements during the last 12 weeks of pregnancy, and to appetite during lactation. Changes in cow live weight in late pregnancy reflected the level of feeding.

Nutritional state, measured as plasma 3-hydroxybutyrate concentrations was closely related to energy intake. On the lower levels of feeding a very severe degree of undernourishment in late pregnancy was indicated but the largest difference in birth weight between groups in the two years was 15%. In cattle, neonatal mortality tends to be greatest in calves of above average birth weight and the moderate reductions observed in these experiments, generally around 10%, are unlikely to influence mortality or subsequent performance. In one experiment birth weight appeared to be reduced by either very low or very high levels of feeding.

Level of nutrition during late pregnancy had no effect on the voluntary intakes of cows during the first 16 weeks of lactation. However, the intake of cows which had been most severely undernourished in late pregnancy declined more slowly in the later stages of lactation. Cows which lost most weight in late pregnancy gained weight most rapidly during lactation. The mean gains over 22 weeks were similar in both breeds.

Milk production showed only moderate increases in early lactation. Peak yields were achieved at around 6-8 weeks after calving. Thereafter, yields declined but were still relatively high at 22 weeks. Milk intake of some calves exceeded 15 kg/d during the period of maximum production. There was no significant effect of prepartum energy intake on subsequent milk production.

No effect of prepartum nutrition on calf performance was observed at any stage of lactation in either year. Calf growth rates averaged around 950 g/hd/d and there was a tendency for the calves from the Hereford × Friesian cows to grow rather faster than those from Blue Grey mothers.

These studies provided an opportunity, in collaboration with the Animal Breeding Research Organisation, to assess the influence of energy intake during late pregnancy on immunoglobulin transfer from cow to calf.

The results show that very severe levels of undernourishment reduced immunoglobulin transfer but not to an extent where concentrations in the calf serum were considered unduly low. Concentrations of immunoglobulins in the colostrum of Blue Grey cows and in the serum of their calves were higher than in the Hereford × Friesian cows and calves. There was evidence of a more rapid uptake of colostrum by calves from Hereford × Friesian cows, and it was noted that cows of both genotypes which transferred high immunoglobulin concentrations in one year tended to do so subsequently (Halliday, Russel, Williams and Peart, 1978).

The principal conclusion which emerges from the first experiments is the general lack of any important production effects of a wide range of nutritional states during late pregnancy. It must be emphasised, however, that all the cows were in good body condition at the beginning of late pregnancy and that they were unequivocally well nourished during lactation (Russel, Peart, Eadie, Macdonald and White, 1979).

Phase 2: 1976-78

A second series of experiments begun in 1976 sought information on the effects on production of the interactions between body condition and levels of feeding in both late pregnancy and early lactation. At this point work was initiated with lactating cows and calves at grass. This was achieved by changing to spring calving which also allowed the extension of the nutritional work to examine interactions of body condition and nutrition in pregnancy and early lactation. Since that time each experiment has comprised a pregnancy treatment period (indoors), an early lactation treatment period of 4-13 weeks (also indoors), and a grazing period.

The need to measure both herbage intake and milk production in grazing cows with minimal disturbance to normal daily grazing routines meant that the calf-suckling technique used previously to measure cow milk yield was no longer acceptable. During the period of changing from autumn to spring calving opportunity was taken to develop a method based on the use of intravenous administration of oxytocin followed by machine milking, and to evaluate it in relation to the calf-suckling technique. This was accomplished (Le Du, Macdonald and Peart, 1979) and the oxytocin method has been successfully used over the last three years.

In the first experiment of this series the cows were fed to supply 75% of maternal maintenance energy requirements in pregnancy. After calving they were divided into two groups, balanced for genotype and calving date, and fed rations to support two contrasting levels of milk production. The cows and calves were subsequently turned out to grass in mid-May. The poorer of the early lactation treatments resulted in only slightly lower levels of milk production but greater weight losses. The calves suckling the better fed cows gained more weight than those suckling the less well fed cows (Peart, Russel, Hodgson, Whitelaw and Macdonald, 1978).

All cows and calves grazed as a single herd in a paddock grazing system at a generous daily allowance of herbage dry matter on a good ryegrass sward. Milk yields increased substantially on turn-out, and the increase was greater in those cows less well fed in early lactation. During the grazing period herbage intakes, milk yields and cow and calf weight gains were marginally higher in cows previously on the lower level of early lactation feeding. Calf growth rates at grass were satisfactory (Hodgson, Peart, Russel, Whitelaw and Macdonald, 1978). This experiment indicated clearly that spring calving suckler cows which enter late pregnancy in good condition, but which have access to a plentiful supply of good quality herbage after turn-out, can withstand a prolonged period of undernourishment in late pregnancy and early lactation with little detriment to ultimate calf performance.

Further work was therefore undertaken to study the interactions between the body condition of cows at 12 weeks prepartum, and their nutrition during late pregnancy and early lactation. In the late pregnancy treatment period some cows in poor condition were fed to avoid significant undernourishment and others were severely undernourished. At parturition the cows were assigned to one of two nutrition levels designed to support two contrasting levels of milk production.

The results confirm the earlier finding that severe undernourishment in late pregnancy will not necessarily have an adverse effect on calf birth weight, and also suggest that, within limits, the body condition of the cow at the beginning of late pregnancy is also unlikely to have a significant effect on calf birth weight. However, some Hereford × Friesian cows in the severely undernourished group were removed from the experiment during pregnancy or early lactation as it was thought that continued undernourishment could have damaged their health. It appears that a condition score of 2.0 to 2.5 at the beginning of late pregnancy is necessary to enable the cow to complete pregnancy on a low level of nutrition without excessive depletion of body reserves.

The treatment which provided better nutrition in both pregnancy and early lactation resulted in higher levels of milk production than those treatments which provided better nutrition in only one period. Calves from the cows in better condition in pregnancy tended to grow faster regardless of level of nutrition in early lactation. Cows on the poorer treatment in early lactation lost approximately twice as much weight as those on the higher treatment.

At turn-out to grass cows and calves were assigned to two groups which were grazed as 'leaders' and 'followers' in a rotational paddock grazing system. The previously observed substantial increase in milk production at turn-out was confirmed in both the leader and follower groups, and the increase was greater in cows on the lower plane of nutrition in early lactation. The leader cows produced more milk, gained more weight and their calves grew more quickly than those of the follower group.

In a more recent experiment an attempt has been made to follow up the questions raised by the large increase in milk production after turn-out to grass. Different planes of nutrition were established over the first 10 weeks of lactation. Cows were turned out to grass in the fourteenth week following an increase in plane of nutrition for some of them during the eleventh to fourteenth week. In this experiment calf suckling frequencies were measured during continuous 48-hour periods before and after turn-out to examine the extent to which suckling behavioural changes might be implicated in the milk production response.

Present position

As the programme has developed it has become increasingly clear that the effects of nutrition on production in the suckler cow and its calf are related to the utilisation and replenishment of body reserves. Consequently, work has recently begun on the difficult question of finding means of measuring body composition in live animals. An attempt is currently being made to quantify relationships between various indices of body composition and directly measured amounts of body fat and protein over a range of cow genotypes and body conditions.

Use has been made in the programme to date of blood metabolites as indices of nutritional status and there has been further exploration of their use, in particular the extent to which they may be used, in absolute as opposed to relative terms, as indices of energy status in pregnancy. Studies of nutrition in pregnancy suggest the need to achieve a level of body condition at parturition appropriate to the nutritional environment in lactation. In the hills and uplands this can vary widely as can the opportunities for its manipulation. Consequently there is a need for more knowledge on the effect of the interaction between body condition at parturition and early lactation nutrition on cow body composition change, lactation performance and calf growth rate.

Work to date has shown that greater inputs of food to the cow in early lactation will increase the amount of milk ingested by the calf and, therefore, its growth rate. Thereafter, where calves have access to good quality grass or to creep feed, nutritional manipulation of the cow has a small effect on calf growth rate. Much remains to be learnt, however, about the factors, including behaviour, which influence milk and solid food intake by the calf. There are also unresolved questions concerning the factors affecting the intake of nutrients by the cow and the partition of nutrients amongst live weight, body composition change and milk yield in relation to nutrition in lactation.

Future plans

Future work will include studies with twin-suckling cows, partly because it is clear that single calves do not test the milk producing capacity of the beef cow, but also because it is anticipated that suckler cow systems in the hills and uplands may in the future be able to support a proportion of multiple-suckling cows.

Nutritional work will continue to constitute a major part of the research programme. Evidence from studies outwith the Organisation points to the importance of nutrition in reproduction, particularly in respect of returns to service and hence calving interval. Consideration is currently being given to the question of obtaining useful information on this very important matter within the framework of the projected nutritional studies and this depends on establishing methods of detecting oestrus and embryo loss.

The grazing studies programme has as its objectives the acquisition of information and understanding on which to found management procedures aimed at improving the efficiency of use of hill and upland pastures for cattle production. The programme will be developed to provide an understanding of the grazing management procedures required to provide the nutrients needed for satisfactory reproduction, lactation and growth, and of the way in which variations in sward conditions influence nutrient intake of cows at different stages of the annual cycle of production, and of calves at different stages of growth.

9

Red deer

Contributors: J Eadie, W J Hamilton, J A Milne
Co-ordinator: J M M Cunningham

Introduction

Proposals for a study of the production potential of red deer (*Cervus elaphus*) were submitted to the Agricultural Research Council by the Rowett Research Institute (RRI) in 1967. HFRO became involved in the project following the conference on red deer held at RRI in January 1969 (Bannerman and Blaxter, 1969).

Initial studies at RRI suggested that deer apparently consumed a greater proportion of heather in their diet than did sheep, and that in captivity they could be easily domesticated (Maloiy, 1968). It was considered that, if deer could be appropriately husbanded, they might use hill land resources efficiently and it was agreed that research into basic husbandry procedures was required before any farming system could be developed. At that time the economics of hill sheep farming were marginal, especially in the poorer hill areas where stocking rates were less than one ewe per four hectares and where deer were indigenous. It seemed pertinent to investigate deer farming as a viable alternative in such areas.

In the winter of 1969/70 HFRO and RRI developed a joint plan which received the necessary approval and the project commenced in March 1970 under the aegis of the Strathfinella Improvement Society (HFRO/RRI collaborative project).

Resources

Physical

Two hundred and fifteen hectares of heather-dominant hill land at Glensaugh, ranging in altitude from 200-450 m, was allocated to the project and the Forestry Commission sanctioned the use of forest roads for access to the farm. The perimeter was fenced and the area divided into paddocks for management and grazing control.

Stock

At the outset it was decided to use animals which were as tame as possible and to this end the initial stock was built up from artificially-reared calves taken from Scottish deer forests when about 1-2 days old. Concern has been expressed that removal of a young calf may cause stress in hinds and, consequently, field observations were made. These have shown that losses due to natural causes or removal of calves have only mild and transient effects on hinds (Corrigal and Hamilton, 1977). Grazing behaviour and appetite did not seem to be affected. There was little seeking after the calf; the udder did not attain a size typical of a suckling hind and regressed from the third day after the loss of the calf.

Many of the original artificially-reared hinds (six from 1970, 49 from 1971 and 26 from 1972) are still on the farm and have remained extremely tame. They respond to a familiar call, readily follow the stockman and can be handled through pens with little difficulty.

The adult stock has varied from 100-120 hind equivalents, i.e. 2.14 to 1.79 ha/hind equivalent, but some paddocks have been more intensively stocked.

Fencing and handling

Over the years several types of fences have been tested. Internal fences of 1.5 m in height have proved adequate but boundary fences must be 2.0 m; where hinds with calves are to be driven a bottom net is essential. Glensaugh is an area prone to snow and this has also allowed an assessment of the ability of fences to withstand adverse conditions; in most winters some fences have been completely buried. A body of user experience, including information on costs of different fence types and their maintenance requirements, has been accumulated.

A set of handling facilities was erected at the outset and is probably more complex than would be required for a commercial operation. Some of the basic principles, for example the circular crush, have been shown to work satisfactorily. Experience of a diversity of types and layouts elsewhere in this country and in New Zealand allows pens to be designed to suit the needs of a particular enterprise.

Performance

It became evident at an early stage that ewe milk substitute was preferable to cow milk substitute and a satisfactory system of artificial rearing was developed, broadly incorporating the principles used in bovine calf-rearing techniques. Satisfactory growth rates have been obtained and rates of live-weight increase of calves have been similar to those reared naturally, i.e. approximately 40 kg for stags and 30-35 kg for hinds by the October of their first year.

By mustering the herd five to six times annually many measurements have been made to characterise growth and development, to describe the various physiological states, and to relate some of these to production parameters.

Only a limited number of animals, mainly stags, have been available for slaughter. Under hill conditions stags have reached 70-85 kg live weight by 16 months and, after losing weight during the ensuing winter, eventually reached 95 kg by 2½ years. The additional gain obtained by keeping stags this extra year is not economically justified so the majority have been slaughtered at 16 months. A slaughter unit has been erected on the deer farm and this allows complete collection of carcase components. Stags reared on the hill throughout their life gave a 57% killing out to yield mean saleable carcases of 44 kg. Reseeded land established in 1975 has also been used for yearlings which have reached 90-105 kg live weight at 16 months to yield mean saleable carcases of 56 kg (KO% 58). Tests of the cooking quality and acceptability of the meat from these domesticated deer have shown that it is likely to be suitable for the United Kingdom market.

It has been found that hinds need to reach 62-65 kg live weight by mating time in October when they are 16 months of age in order to breed in their second year. This is rarely achieved by feral animals. In this study calving performance of 90-95% at two years old has been obtained. However, in experiments in which lower levels of first winter nutrition have been imposed levels of performance have been depressed to the extent that some groups failed to breed. Thereafter, apart from periods of storm feeding, which can be prolonged as in 1978/79, no feeding has been practised in winter and inputs for several years were minimal. Supplementary feeding (0.5 kg/hd/d) has been offered in May and June during late pregnancy and early lactation.

The stag-to-hind ratio has been examined; given adequate size (a function of age) and libido a stag will cover up to 30 hinds, all calving in June/early July. Although individual stag mating groups have been used for experimental purposes it would be uneconomic in practice to provide many paddocks for this purpose so several stag/hind groups have to be run together. This could give rise to the conflicts customarily observed in the wild. However, by isolating stags for 5-10 days with their harem and selecting stags close in the dominance order, subsequent fighting can be virtually eliminated with three groups in a paddock of around 50 ha.

Health

Throughout the experiment the health status of all animals has been regularly monitored and most fatalities have been subjected to post-mortem examination. Experience shows that the deer is an extremely healthy animal with few attendant disease problems when run in hill conditions. The most common and troublesome parasite has been the lungworm (*Dictyocaulus viviparus*) causing parasitic bronchitis, and experiments in the control of the disease are in progress. Routine control of warble fly (*Hypoderma diana*) has been shown to be desirable and appropriate preparations identified, as not all commercially available products are suitable for use with deer. The head fly (*Hydrotaea irritans*) is a particular pest of stags in velvet and some collaborative work was undertaken to further the knowledge of the basic biology of the fly and its life cycle, particularly to establish the breeding sites and sites of overwintering. Although other diseases and pests have been identified none has given rise to serious losses. Another problem which can arise with the intensive handling of semi-tame animals is limb fracture or other injury.

Related studies

Each of the collaborating institutes has been responsible for specific areas of more intensive investigation. HFRO has been concerned with detailed studies of the digestive physiology of deer in relation to low quality diets. There is a need to examine whether there are differences between sheep and deer in their ability to utilise hill land resources for meat production. Differences between the species could arise because of differences in grazing behaviour, in the ability to consume and digest food, in the efficiency of utilisation of the end products of digestion and in nutrient requirements.

The second aspect referred to above was examined in an experiment where deer and Scottish Blackface sheep were compared when given poor quality *Agrostis-Festuca* and heather herbage in winter and dried grass diets in summer (Milne, MacRae, Spence and Wilson, 1976, 1978). When compared per unit of live weight deer and sheep ate similar quantities of high quality dried grass in summer. Sheep digested dried grass better than did deer and this was associated with a longer mean retention time of undigested residues in the gut of the sheep.

With the poor quality diets in winter a different picture emerged. The intakes of *Agrostis-Festuca* and heather herbages by deer were twice that of sheep. Moreover, whilst intake of these herbages by sheep increased by 15% between January and April, intake by deer expanded by 70%. The sheep digested the *Agrostis-Festuca* better than did deer but deer digested heather slightly better than did sheep. Retention times of undigested residues in the gut of sheep were greater than those of deer with both herbages. Since the digestibility and mean retention times in deer did not change with the seasonal increase in intake, this indicated that the weight or volume in the digestive tract had increased between January and April. This was confirmed in a subsequent experiment (Milne and Spence, unpublished) where it was shown that the rumen contained a greater volume and weight of digesta in April than in January. These experiments demonstrated that intakes of digestible organic matter from hill vegetation in winter by deer were considerably higher than those by sheep. However, there is still insufficient comparative information on other factors which may influence performance, although these are being investigated by other collaborating institutes, to enable firm conclusions to be drawn about whether red deer are better adapted than sheep to hill environments in winter.

HFRO has also been concerned with assessing the tractability of artificially- and naturally-reared hinds, monitoring the effects of grazing pressure on the hill paddocks and studying grazing behaviour. Because some hinds are extremely tame it was possible to record every bite taken by two hinds from dawn to dusk and to note the plant species eaten. Lactating hinds select a diet containing more grass than dry hinds and they also spend considerably more time grazing.

Enough is now known about the husbandry of red deer to begin to examine production on a systems basis, and this will be the main objective of studies in the years ahead. The first report on this joint venture was published in 1974 by HMSO (Blaxter, Kay, Sharman, Cunningham and Hamilton, 1974). The second report which will cover the entire period up to 1978 is now in preparation.

References

BANNERMAN, M M and BLAXTER, K L Eds (1969). The husbanding of red deer. Proceedings of a conference held at the Rowett Institute, Aberdeen in January 1969. Aberdeen, HIDB/RRI, 1969.

MALOIY, G M O (1968). The physiology of digestion and metabolism in the red deer *(Cervus elaphus L.). PhD Thesis, University of Aberdeen.*

Publications

The following is a complete alphabetical list of published papers from HFRO: joint authors from outwith the organisation are identified by an asterisk. The full citation to external papers referred to in the text can be found at the end of each chapter.

ALLEN*, S.E., CARLISLE*, A., WHITE*, E.J. and EVANS, C.C. (1968). The plant nutrient content of rainwater. *J. Ecol.* **56**, 497-504.

ALLEN*, S.E., EVANS, C.C. and GRIMSHAW*, H.M. (1969). The distribution of mineral nutrients in soil after heather burning. *Oikos,* **20**, 16-25.

ARMAN*, P., HAMILTON, W.J. and SHARMAN*, G.E.M. (1978). Observations on calving of free ranging tame red deer *(Cervus elaphus). J. Reprod. Fert.* **54**, 279-283.

ARMSTRONG, Richard H. and EADIE, J. (1973a). Lamb growth on grass and clover diets. *Proc. Br. Soc. Anim. Prod.,* **2, 68**. (Abstract).

ARMSTRONG, Richard H. and EADIE, J. (1973b). Some aspects of the growth of hill lambs. *HFRO 6th Report,* 1971-73, 57-68.

ARMSTRONG, Richard H. and EADIE, J. (1977). The growth of hill lambs on herbage diets. *J. Agric. Sci., Camb.* **88**, 683-692.

ARMSTRONG, Robin H. (1973). *Report of a visit to Australia and New Zealand, in October-November 1972, to study sheep and cattle production on hill and high ground.* U.K. Farming Scholarships Trust.

ARMSTRONG, Robin H. (1975). Fencing and the hill farmer. *Luing J.* 55-59.

ARMSTRONG, Robin H. and CAMERON*, A.E. (1959). Hexoestrol implantation of Blackface wether lambs. *Anim. Prod.,* **1,** 37-40.

ARMSTRONG, Robin H. and CAMERON*, A.E. (1961). Further studies of hexoestrol implantation of Blackface wether lambs. *Anim. Prod.,* **3**, 295-300.

ARMSTRONG, Robin H., EADIE, J. and MAXWELL, T.J. (1978). The development and assessment of a modified hill sheep production system at Sourhope, in the Cheviot Hills (1968-1976). *HFRO 7th Report,* 1974-77, 69-101.

ATTWOOD*, P.R. and HUNTER, R.F. (1957). A method for studying the preferential grazing of hill sheep. *Br. J. Anim. Behav.,* **5**, 149-152.

BEEVER*, D.E., KELLAWAY*, R.C., THOMSON*, D.J., MACRAE, J.C., EVANS, C.C. and WALLACE*, A.S. (1978). A comparison of two non-radioactive digesta marker systems for the measurement of nutrient flow at the proximal duodenum of calves. *J. Agric. Sci., Camb.,* **90**, 157-163.

BLACK, J.S. (1967). The digestibility of indigenous hill pasture species. *HFRO 4th Report,* 1964-67, 33-37.

BLAXTER*, K.L., KAY*, R.N.B., SHARMAN*, G.A.M., CUNNINGHAM, J.M.M. and HAMILTON, W.J. (1974). *Farming the red deer.* The first report of an investigation by the Rowett Research Institute and the Hill Farming Research Organisation. Department of Agriculture and Fisheries for Scotland, Edinburgh, HMSO.

BOGGIE*, R., HUNTER, R.F. and KNIGHT*, A.H. (1958). Studies of the root development of plants in the field using radioactive tracers. Pt. I Communities growing in mineral soil. Pt.II Communities growing in deep peat. *J. Ecol.,* **46**, 621-639.

Red deer stag

Suckler cow   with calves

Boyd*, J.M., Doney, J.M., Gunn, R.G. and Jewell*, P.A. (1964). The Soay sheep of the island of Hirta, St. Kilda. A study of a feral population. *Proc. Zool. Soc., Lond.*, **142**, 129-163.

Chambers, A.R.M., Milne, J.A., Russel, A.J.F. and White, I.R. (1975). An apparatus for the measurement and sampling of urine from grazing female sheep. *Proc. Nutr. Soc.*, **34**, 71A-72A. (Abstract).

Chambers, A.R.M. and Skedd, E. (1975). Measurement of tritiated water in blood plasma (a short note). *Newsletter on the Application of Nuclear Methods in Biology and Agriculture*, **5**, 17.

Chambers, A.R.M., White, I.R., Russel, A.J.F. and Milne, J.A. (1976). An instrument for sampling and measuring the volume output of urine from grazing female sheep. *Med. Biol. Eng.*, **14**, 665-670.

Christie, A. (1978). *The nutritive value of heather (Calluna vulgaris L. Hull) to hill sheep.* M.Phil. Thesis, University of Edinburgh.

Copeman*, G.J.F. and Nicholson, I.A. (1958). Autumn and winter grazing in northern districts. *Agric. Rev., Lond.*, **3**, 22-30.

Corrigal*, W. and Hamilton, W.J. (1977). Reaction of red deer (*Cervus elaphus* L.) hinds to removal of their suckled calves for hand rearing. *J. Appl. Anim. Ethol.*, **3**, 47-55.

Cunningham, J.M.M. (1971a). Developments in hill farming. *Scott. For.*, **25**, 175-180.

Cunningham, J.M.M. (1971b). Hill farming industry. *Proc. Nutr. Soc.*, **30**, 195-197.

Cunningham, J.M.M. (1972a). Developments in hill farming. *Soc. of Foresters Meeting*, Carlisle, 1971.

Cunningham, J.M.M. (1972b). Improvement of hill sheep production. *Peter Wilson Bequest Course*, Royal (Dick) School of Veterinary Studies, Edinburgh.

Cunningham, J.M.M. (1972c). Improvement of mountain hill sheep production in Great Britain. *Inter. Symp. Semi-Mountain and Mountain Sheep Breeding*, Varna, Bulgaria, 26.

Cunningham, J.M.M. (1973). Agriculture as the primary object of land use. *Forestry* Supplement, 7-10.

Cunningham, J.M.M. (1976). Scottish food production: hill farming. *Chemy. Ind.*, 17 July, 599-602.

Cunningham, J.M.M. (1978). Hill sheep: production problems. *Vet. Rec.*, **103**, 177. (Abstract).

Cunningham, J.M.M. (1979). Land use for ruminants — northern hill areas. CEC Seminar *'The future in the EEC of beef production'*, Abano-Terme, Italy, Nov. 1978. (In press).

Cunningham, J.M.M. and Eadie, J. (1976a). Science to the aid of the hill sheep farmer in Great Britain. Suppl. to: *Proc. 11th Biennial Conference of Aust. Soc. Anim. Prod.*, Adelaide, 9-13 February, 1976.

Cunningham, J.M.M. and Eadie, J. (1976b). A visit to . . . Nepal . . . Mongolia. *Blackface J.*, **28**, 42-43.

CUNNINGHAM, J.M.M., EADIE, J., MAXWELL, T.J. and SIBBALD, A.R. (1978). Inter-relations between agriculture and forestry: an agricultural view. *Scott. For.*, **32**, 182-193.

CUNNINGHAM, J.M.M. and MAXWELL, T.J. (1979). Improved sheep production on hill farms. *Vet. Annual*, **19**, 69-73.

CUNNINGHAM, J.M.M. and RUSSEL, A.J.F. (1979). The technical development of sheep production from hill land in Great Britain. *Livest. Prod. Sci.* (In press).

CUNNINGHAM, J.M.M. and SMITH, A.D.M. (1978). The impact of technical advance in hill and upland cattle systems. In: *The future of upland Britain*, ed. by R.B. Tranter. Centre for Agricultural Strategy Paper 2, 34-45.

CUNNINGHAM, J.M.M., SMITH, A.D.M. and DONEY, J.M. (1970). Trends in livestock populations in hill areas in Scotland. *HFRO 5th Report*, 1967-70, 88-95.

CUNNINGHAM, J.M.M., WATT*, J.A.A. and GIBSON*, T.E. (1973). Management systems and productivity of hill and upland sheep. *Vet. Rec.*, **92**, 581-584.

DAVIES, G.E. HUNTER, R.F. and KING, J. (1960). Hill pasture improvement with Dalapon. *5th Br. Weed Control Conf. Proc.*, **1**, 157-164.

DAVIES, G.E., NEWBOULD, P. and BAILLIE, G.J. (1979). The effect of controlling bracken *(Pteridium aquilinum (L) Kuhn)* on pasture production. *Grass and Forage Sci.* (In press).

DONEY, J.M. (1962). Inbreeding in Blackface sheep. *Anim. Prod.*, **4**, 293. (Abstract).

DONEY, J.M. (1963a). The effects of exposure in Blackface sheep with particular reference to the role of the fleece. *J. Agric. Sci.*, **60**, 267-273.

DONEY, J.M. (1963b). The fleece of the Scottish Blackface sheep. III. The relative importance of the fleece components. *J. Agric. Sci.*, **61**, 45-53.

DONEY, J.M. (1964a). The fleece of the Scottish Blackface sheep. IV. The effects of pregnancy, lactation and nutrition on seasonal wool production. *J. Agric. Sci.*, **62**, 59-66.

DONEY, J.M. (1964b). Genetic effects on feed intake and efficiency of feed utilisation. *Anim. Prod.*, **6**, 262 (summary).

DONEY, J.M. (1966a). Inbreeding depression in grazing Blackface sheep. *Anim. Prod.*, **8**, 261-266.

DONEY, J.M. (1966b). Breed differences in response of wool growth to annual nutritional and climatic cycles. *J. Agric. Sci.*, **67**, 25-30.

DONEY, J.M. (1967a). The effect of inbreeding on feed consumption and utilisation by sheep. *Anim. Prod.*, **9**, 359-364.

DONEY, J.M. (1967b). Nutrition and wool growth. *HFRO 4th Report*, 1964-67, 77-84.

DONEY, J.M. (1967c). Production of protein fibres. *Rev. Text. Prog.*, **17**, 58-80.

DONEY, J.M. (1970). [Problems of hill sheep production in Scottish agriculture.] *Ovcactvi,. roc* **IV** c.6, 7-16. (Czech.)

DONEY, J.M. (1979). Nutrition and the reproductive function in female sheep. British Council Course 729: *Management and diseases of sheep*, Edinburgh, March 1978. (In press).

DONEY, J.M. and EADIE, J. (1967). The wool growth response to the annual cycle of grazing intake in Cheviot wethers. *J. Agric. Sci., Camb.,* **69,** 411-416.

DONEY, J.M. and EVANS, C.C. (1968). The influence of season and nutrition on the sulphur content of wool from Merino and Cheviot sheep. *J. Agric. Sci., Camb.,* **70,** 111-116.

DONEY, J.M. and EVANS, C.C. (1970). The influence of nutrition during winter on growth rate and sulphur content of wool of pregnant Scottish Blackface and Romney ewes. *J. Agric. Sci., Camb.,* **75,** 133-136.

DONEY, J.M. and GRIFFITHS, J.G. (1967). Wool growth regulation by local skin cooling. *Anim. Prod.,* **9,** 393-397.

DONEY, J.M. and GUNN, R.G. (1972). The effect of weather at mating time on reproductive performance of ewes. *Proc. VII Inter. Congress on Animal Production and Artificial Insemination,* Munich, **3,** 2056-2060.

DONEY, J.M. and GUNN, R.G. (1973). Progress in studies on the reproductive performance of hill sheep. *HFRO 6th Report,* 1971-73, 69-73.

DONEY, J.M. and GUNN, R.G. (1976). Factors affecting reproductive performance in sheep. *Vet. Annual,* **16,** 70-74.

DONEY, J.M. and GUNN, R.G. (1979). Environmental effects on reproductive performance of female sheep. In: *Environmental factors in mammalian reproduction. Vol. IV. Biology and environment.* London, Macmillan. (In press).

DONEY, J.M. GUNN, R.G. and GRIFFITHS, J.G. (1973). The effects of premating stress on the onset of oestrus and on ovulation rate in Scottish BF ewes. *J. Reprod. Fert.,* **35,** 381-384.

DONEY, J.M., GUNN, R.G. and SMITH, W.F. (1973). Transuterine migration and embryo survival in sheep. *J. Reprod. Fert.,* **34,** 363-367.

DONEY, J.M., GUNN, R.G., SMITH, W.F. and CARR*, W.R. (1976). Effects of premating environmental stress, ACTH, cortisone acetate or metyrapone on oestrus and ovulation in sheep. *J. Agric. Sci., Camb.,* **87,** 127-132.

DONEY, J.M. LOUDA*, F. and KRIZEK*, J. (1978). [Determination of milk production of the Merino ewes by two measurement methods.] *Zivocisna Vyroba,* **23,** 909-914. (Czech.)

DONEY, J.M., LOUDA*, F., SMERHA*, J. and SKRIVAN*, M. (1977). [The annual cycle of sperm production in rams.] *Zivocisna Vyroba,* **22,** 923-929. (Czech)

DONEY, J.M. and MUNRO, Joan (1962). The effect of suckling, management and season on sheep milk production as estimated by lamb growth. *Anim. Prod.,* **4,** 215-220.

DONEY, J.M. and PEART, J.N. (1976). The effect of sustained lactation on the intake of solid foods and growth rate of lambs. *J. Agric. Sci., Camb.,* **87,** 511-518.

DONEY, J.M., PEART, J.N., SMITH, W.F. and LOUDA*, F. (1979). A consideration of the techniques for estimation of milk yield by suckled sheep and a comparison of estimates obtained by two methods in relation to the effect of breed, level of production and stage of lactation. *J. Agric. Sci., Camb.,* **92,** 123-132.

DONEY, J.M. and RUSSEL, A.J.F. (1968). Differences amongst breeds of sheep in food requirements for maintenance and live-weight change. *J. Agric. Sci., Camb.,* **71,** 343-349.

DONEY, J.M. and RUSSEL, A.J.F. (1969). Experimental estimation of maintenance requirements and their use in the annual production cycle of the ewe. *VIIIth International Nutrition Congress, Prague.*

DONEY, J.M., RYDER*, M.L., GUNN, R.G. and GRUBB*, P. (1974). Colour, conformation, affinities, fleece and patterns of inheritance of the Soay sheep. In: *Island survivors, the ecology of the Soay sheep of St. Kilda,* ed. by P.A. Jewell, C. Milner and J. Morton Boyd, London, Athlone Press., 88-125.

DONEY, J.M. and SLEE*, J. (1972). General meteorological effects on sheep. In: *Progress in Biometeorology,* **Vol. I,** ed. by S.W. Tromp. Amsterdam. Swets & Zeitlinger.

DONEY, J.M. and SMITH, W.F. (1961a). The fleece of the Scottish Blackface sheep. I. Seasonal changes in wool production and fleece structure. *J. Agric. Sci.,* **56,** 365-374.

DONEY, J.M. and SMITH, W.F. (1961b). The fleece of the Scottish Blackface sheep. II. Variation in fleece components over the body of the sheep. *J. Agric. Sci.,* **56,** 375-378.

DONEY, J.M. and SMITH, W.F. (1962). Multiple medullae in wool fibres. *Nature, Lond.,* **195,** 723.

DONEY, J.M. and SMITH, W.F. (1964). Modification of fleece development in Blackface sheep by pre- and post-natal nutrition. *Anim. Prod.,* **6,** 155-167.

DONEY, J.M. and SMITH, W.F. (1966). Fleece characteristics of three strains of Scottish Blackface sheep. *Anim. Prod.,* **8,** 313-320.

DONEY, J.M. and SMITH, W.F. (1968). Infertility in inbred ewes. *J. Reprod. Fert.,* **15,** 277-282.

DONEY, J.M. and SMITH, W.F. (1969). Casting of the fleece in sheep. Estimation of experimentally induced fibre shedding rate. *J. Agric. Sci., Camb.,* **73,** 231-237.

DONEY, J.M., SMITH, W.F. and GUNN, R.G. (1976). Effects of post-mating environmental stress or administration of ACTH on early embryonic loss in sheep. *J. Agric. Sci., Camb.,* **87,** 133-136.

DUCKWORTH*, J., BENZIE*, D., CRESSWELL*, E., HILL*, R. and DALGARNO*, A.C. with ROBINSON, J.F. and ROBSON*, H.W. (1961). Dental malocclusion and rickets in sheep. *Res. Vet. Sci.,* **2,** 375-380.

The late DUCKWORTH*, J., HILL*, R., BENZIE*, D. and DALGARNO*, A.C. in collaboration with ROBINSON, J.F. (1962). Studies of the dentition of sheep. I. Clinical observations from investigations into the shedding of permanent incisor teeth by hill sheep. *Res. Vet. Sci.,* **3,** 1-17.

EADIE, J. (1967a). Improvement of grass production and utilisation on natural grassland in different climates. *J. Br. Grassld. Soc.,* **22,** 21-25.

EADIE, J. (1967b). The nutrition of grazing hill sheep; utilisation of hill pastures. *HFRO 4th Report,* 1964-67, 38-45.

EADIE, J. (1969). Systems of hill sheep production. *Proc. 3rd Ann. Conf. Reading Univ. Agric. Club.*

EADIE, J. (1970a). Hill sheep production systems development. *HFRO 5th Report,* 1967-70, 70-87.

EADIE, J. (1970b). Sheep production and pastoral resources. In: *Animal Populations in Relation to their Food Resources*. Br. Ecol. Soc. Symp. No. 10, Oxford, Blackwell, 7-24.

EADIE, J. (1971). Hill pastoral resources and sheep production. *Proc. Nutr. Soc.*, **30**, 204-210.

EADIE, J. (1973). Sheep production systems development on the.hills. In: *Hill pasture improvement and its economic utilisation*, Colloquium Proceedings No. 3 Potassium Institute Ltd., Edinburgh, 1972, 131-137.

EADIE, J. (1974). Biological efficiency in hill sheep production. *J. Sci. Fd. Agric.*, **25**, 341. (Abstract)

EADIE, J. (1976). Systems of utilisation of hill grazings. In: *Pasture utilisation by the grazing animal*. Br. Grassld Soc. Occ. Symp. **8**, 135-139.

EADIE, J. (1978). Increasing output in hill sheep farming. *J. Roy. Agric. Soc. England*, **139**, 103-114.

EADIE, J. (1979a). Extensive sheep systems in Great Britain. British Council Course 729: *Management and Diseases of Sheep*, Edinburgh, March 1978. (In press)

EADIE, J. (1979b). Hill sheep farming systems in Great Britain. *Int. Hill Land Symposium*, Morgantown, West Virginia, 3-9 October, 1976. (In press)

EADIE, J., ARMSTRONG, Robin H. and MAXWELL, T.J. (1973). Hill sheep production systems: development in the Cheviots. In: *Hill pasture improvement and its economic utilisation*, Colloquium Proceedings No. 3 Potassium Institute Ltd., Edinburgh, 1972, 139-144.

EADIE, J., ARMSTRONG, Robin H., MAXWELL, T.J. and CURRIE, D.C. (1979). Responses in output achieved in improved systems of sheep production in two hill environments. I. In the Eastern Cheviots. II. On blanket peat in the West of Scotland. *Int. Hill Land Symposium*, Morgantown, West Virginia, 3-9 October, 1976. (In press)

EADIE, J. and BLACK, J.S. (1968). Herbage utilisation on hill pastures. In: *Hill-land Productivity*. Br. Grassld Soc. Occ. Symp., **4**, 191-195.

EADIE, J. and CUNNINGHAM, J.M.M. (1969). The efficiency of hill sheep production systems. *Proc. Conf. on Crop Potential*, Univ. College Wales, 239-249.

EADIE, J., GRANT, Sheila A. *et al.* (Contributors) (1977). A guide to good muirburn practice. Muirburn Working Party. DAFS and NCC. Edinburgh, HMSO.

EADIE, J. and MAXWELL, T.J. (1974). Economic improvement of hill productivity. *Norgrass*, **14**, 23-30.

EADIE, J. and MAXWELL, T.J. (1975). Systems research in hill sheep farming. In: *Study of Agricultural Systems*, ed. by G.E. Dalton. London, Applied Science Publishers, 395-413.

EADIE, J., MAXWELL, T.J., KERR, C.D. and CURRIE, D.C. (1973). Hill sheep production systems; development on blanket peat. In: *Hill pasture improvement and its economic utilisation*, Colloquium Proceedings No. 3 Potassium Institute Ltd., Edinburgh, 1972, 145-153.

EADIE, J. and SMITH, A.D.M. (1978). The impact of technical advance on hill and upland sheep production. In: *The future of upland Britain*, ed. by R.B. Tranter. Centre for Agricultural Strategy Paper 2, 22-33.

EGAN*, A.R. and MACRAE, J.C. (1978). The contribution of threonine to the glucose economy of hill sheep. *Proc. Nutr. Soc.*, **37**, 15A. (Abstract)

EVANS, C.C. (1970). X-ray fluorescence analysis of light elements in plant and faecal materials. *Analyst, Lond.*, **95**, 919-929.

EVANS, C.C. (1974). X-ray fluorescence spectrometry. In: *Chemical analysis of ecological materials*, ed. by S.E. Allen. Oxford, Blackwells, 404-412.

EVANS, C.C. and ALLEN*, S.E. (1971). Nutrient losses in smoke produced during heather burning. *Oikos*, **22**, 149-154.

EVANS, C.C. and GRIMSHAW*, H.M. (1968). Use of lanthanum and sulphuric acid to suppress interferences in the flame photometric determination of calcium in soil extracts. *Talanta*, **15**, 413-415.

EVANS, C.C., MACRAE, J.C. and WILSON, S. (1977). Determination of ruthenium and chromium by X-ray fluorescence spectrometry and the use of inert ruthenium (II) phenanthroline as a solid phase marker in sheep digestion studies. *J. Agric. Sci., Camb.*, **89**, 17-22.

FIELD*, A.C., SUTTLE*, N.F. and GUNN, R.G. (1968). Seasonal changes in the composition and mineral content of the body of hill ewes. *J. Agric. Sci., Camb.*, **71**, 303-310.

FIELD*, A.C., SYKES*, A.R. and GUNN, R.G. (1974). Effects of age and state of incisor dentition on faecal output of dry matter, and on faecal and urinary output of nitrogen and minerals, of sheep grazing hill pastures. *J. Agric. Sci., Camb.*, **83**, 151-160.

FLOATE, M.J.S. (1967). Hill soils. *HFRO 4th Report*, 1964-67. 17-21.

FLOATE, M.J.S. (1968). The nutrient cycle in hill soils with particular reference to the mineralisation of carbon, nitrogen and phosphorus from plant litter and sheep excreta. In: *Hill-land Productivity*. Br. Grassld Soc. Occ. Symp., **4**, 149-155.

FLOATE, M.J.S. (1972a). Decomposition of organic materials from hill soils and pastures. II. Comparative studies on the mineralisation of carbon, nitrogen and phosphorus from plant materials and sheep faeces. *Soil Biol. Biochem.*, **2**, 173-185.

FLOATE, M.J.S. (1970b). Decomposition of organic materials from hill soils and pastures. III. The effect of temperature on the mineralisation of carbon, nitrogen and phosphorus from plant materials and sheep faeces. *Soil Biol. Biochem.*, **2**, 187-196.

FLOATE, M.J.S. (1970c). Decomposition of organic materials from hill soils and pastures. IV. The effects of moisture content on the mineralisation of carbon, nitrogen and phosphorus from plant materials and sheep faeces. *Soil Biol. Biochem.*, **2**, 275-283.

FLOATE, M.J.S. (1970d). Mineralisation of nitrogen and phosphorus from organic materials of plant and animal origin and its significance in the nutrient cycle in grazed upland and hill soils. *J. Br. Grassld Soc.*, **25**, 295-302.

FLOATE, M.J.S. (1970e). Plant nutrient cycling in hill land. *HFRO 5th Report*, 1967-70, 15-34.

FLOATE, M.J.S. (1971). The significance in hill pasture improvement of the recirculation of plant nutrients in sheep excreta, and of increased pasture utilisation. In: *Comparison between natural and artificial grassland*, Proc. 4th General Meeting, Eur. Grassld Fed., Lausanne, Switzerland, 61-68.

FLOATE, M.J.S. (1972a). Nutrient circulation in the Soil-Plant-Animal System. *Proc. North of England Soil Discussion Group* No. 7, 11-27.

FLOATE, M.J.S. (1972b). Geology, geomorphology and soils of Scottish Borders. *Proc. North of England Soil Discussion Group* No. 7, 35-39.

FLOATE, M.J.S. (1973). Soils associated with grass and heather moorland and some long term changes related to grazing influences. *J. Sci. Fd Agric.*, **24**, 1149. (Abstract)

FLOATE, M.J.S. (1977a). British hill soil problems. *Soil Sci.*, **123**, 325-331.

FLOATE, M.J.S. (1977b). Changes in the soil nutrient pools, a contribution to Chapter 7 in: *The cycling of mineral nutrients in agricultural ecosystems. Agro-Ecosystems*, **4**, 292-295.

FLOATE, M.J.S. (1977c). The control of nutrient cycling, Chapter 3 in: *The cycling of mineral nutrients in agricultural ecosystems. Agro-ecosystems*, **4**, 7-16.

FLOATE, M.J.S. ed, (1978a). Hill and upland grassland improvement. *ARC Res. Rev.*, **4**, 11-14.

FLOATE, M.J.S. (1978b). The reclamation of acid hill soils in Britain. *11th Int. Congress, Int. Soc. Soil Sci.* Commission II (Soil Chemistry). Section 6. Chemistry of soil reclamation, Vol. **I**, 365-366. (Abstract)

FLOATE, M.J.S. (1979). Dynamics of nitrogen recycling in grassland ecosystems. *Agro-Ecosystems.* (Letter) (In press)

FLOATE, M.J.S., EADIE, J., BLACK, J.S. and NICHOLSON, I.A. (1973). Improvement of *Nardus* dominant hill pasture by grazing control and surface treatment and its economic assessment. In: *Hill pasture improvement and its economic utilisation.* Colloquium Proceedings No. 3 Potassium Institute Ltd., Edinburgh, 1972, 33-39.

FLOATE, M.J.S. and KING, J. (1974). Improvement of hill grazings. *Norgrass*, **14**, 17-22.

FLOATE, M.J.S. and PIMPLASKAR, M.S. (1976). Some anomalous results for available-P in hill soils, and experimental attempts towards their resolution. *J. Sci. Fd Agric.*, **27**, 591. (Abstract)

FLOATE, M.J.S. and TORRANCE*, C.J.S. (1970). Decomposition of the organic materials from hill soils and pastures. I. Incubation method for studying the mineralisation of carbon, nitrogen and phosphorus. *J. Sci. Fd Agric.*, **21**, 116-120.

FOOT, Janet Z. (1972). A note on the effect of body condition on the voluntary intake of dried grass wafers by Scottish Blackface ewes. *Anim. Prod.*, **14**, 131-134.

FOOT, Janet Z., DONEY, J.M., MAXWELL, T.J. and GUNN, R.G. (1978). A comparison of body- and carcass-composition of Scottish Blackface, Border Leicester × Blackface and Texel × Blackface lambs. *Anim. Prod.*, **26**, 371. (Abstract)

FOOT, Janet Z. and RUSSEL, A.J.F. (1971). The effect of group-feeding on the food requirements of Blackface ewes. *Proc. Nutr. Soc.*, **30**, 45A. (Abstract)

Foot, Janet Z. and Russel, A.J.F. (1973). Some nutritional implications of group-feeding hill sheep. *Anim. Prod.*, **16**, 293-302.

Foot, Janet Z. and Russel, A.J.F. (1978). Pattern of intake of three roughage diets by non-pregnant, non-lactating Scottish Blackface ewes over a long period, and the effects of previous nutritional history on current intake. *Anim. Prod.*, **26**, 203-215.

Foot, Janet Z. and Russel, A.J.F. (1979). The relationship in ewes between voluntary intake during pregnancy and forage intake in lactation and after weaning. *Anim. Prod.*, **28**, 25-39.

Foot, Janet Z., Russel, A.J.F., Maxwell, T.J. and Morris, P. (1973). Variation in intake among group-fed pregnant Scottish Blackface ewes given restricted amounts of food. *Anim. Prod.*, **17**, 169-177 and *Proc. Nutr. Soc.*, **32**, 26A. (Abstract)

Foot, Janet Z., Skedd, E. and McFarlane, D.N. (1979). Body composition in lactating sheep and its indirect measurement in the live animal using tritiated water. *J. Agric. Sci., Camb.*, **92**, 69-81.

Foot, Janet Z. and Tissier*, M. (1977). Quantites d'aliments ingeres par les brebis pendant la lactation. *Bull. Techn. CRZV Theix-INRA*, **30**, 81. (Abstract)

Foot, Janet Z. and Tulloh*, N.H. (1977). Effects of two paths of live-weight change on the efficiency of feed use and on body composition of Angus Steers. *J. Agric. Sci., Camb.*, **88**, 135-142.

Foster, W.N.M. (1967). Tick-borne fever and tick pyaemia. *HFRO 4th Report, 1964-67*, 91-95.

Foster, W.N.M. (1968). The effect of dipping on tick infestation in young lambs. *Vet. Rec.*, **82**, 348-351.

Foster, W.N.M. and Cameron, A.E. (1968a). Aetiology of enzootic staphylococcal infection (tick pyaemia) in lambs. Field investigations into the relationship between tick-borne fever and tick pyaemia. *J. Comp. Path.*, **78**, 243-250.

Foster, W.N.M. and Cameron, A.E. (1968b). Thrombocytopenia in sheep associated with experimental tick-borne fever infection. *J. Comp. Path.*, **78**, 251-254.

Foster, W.N.M. and Cameron, A.E. (1970a). Observations on the functional integrity of neutrophil leucocytes infected with tick-borne fever. *J. Comp. Path.*, **80**, 487-491.

Foster, W.N.M. and Cameron, A.E. (1970b). Observations on ovine strains of tick-borne fever. *J. Comp. Path.*, **80**, 429-436.

Foster, W.N.M., Foggie*, A. and Nisbet*, D.I. (1968). Haemorrhagic enteritis in sheep experimentally infected with tick-borne fever. *J. Comp. Path.*, **78**, 255-258.

Gething*, P.A., Newbould, P. and Patterson*, J.B.E. (eds.) (1973). *Hill pasture improvement and its economic utilisation.* Colloquium Proceedings No. 3 of Potassium Institute Ltd., Edinburgh, 1972. Henley-on-Thames, Oxon.

Grant, Sheila A. (1968a). Heather regeneration following burning: A Survey. *J. Br. Grassld Soc.*, **23**, 26-33.

GRANT, Sheila A. (1968b). Temperature and light as factors limiting the growth of hill pasture species. In: *Hill-land Productivity*. Br. Grassld Soc. Occ. Symp., **4**, 1-5.

GRANT, Sheila A. (1971a). The measurement of primary production and utilisation on heather moors. *J. Br. Grassld Soc.*, **26**, 51-58.

GRANT, Sheila A. (1971b). Interactions of grazing and burning on heather moors. 2. Effects on primary production and level of utilisation. *J. Br. Grassld Soc.*, **26**, 173-181.

GRANT, Sheila A., BARTHRAM, G.T., LAMB, W.I.C. and MILNE, J.A. (1978). Effect of season and level of grazing on the utilisation of heather by sheep. 1. Responses of the sward. *J. Br. Grassld Soc.*, **33**, 289-300.

GRANT, Sheila A. and CAMPBELL, D.R. (1978). Seasonal variation in the *in vitro* digestibility and structural carbohydrate content of some commonly grazed plants of blanket bog. *J. Br. Grassld Soc.*, **33**, 167-173.

GRANT, Sheila A. and HUNTER, R.F. (1962). Ecotypic differentiation in *Calluna vulgaris*. *New Phytol.*, **61**, 44-55.

GRANT, Sheila A. and HUNTER, R.F. (1966). The effects of frequency and season of clipping on the morphology, productivity and chemical composition of *Calluna vulgaris* (L) Hull. *New Phytol.*, **65**, 125-133.

GRANT, Sheila A. and HUNTER, R.F. (1968a). Interactions of grazing and burning on heather moors and their implications in heather management. *J. Br. Grassld Soc.*, **23**, 285-293.

GRANT, Sheila A. and HUNTER, R.F. (1968b). Variation in yield, maturity type, winter greenness and sensitivity to cutting of hill grass species. *J. Br. Grassld Soc.*, **23**, 149-155.

GRANT, Sheila A., HUNTER, R.F. and CROSS, C. (1963). The effects of muirburning *Molinia*-dominant communities. *J. Br. Grassld Soc.*, **18**, 249-257.

GRANT, Sheila A. and KING, J. (1969). Altitude and Grass Growth — Some observations on apparent exposure effects. *Third Symp. on Shelter Research, Cambridge*. London, MAFF Land Improvement Division, 77-86.

GRANT, Sheila A. and KING, J. (1970). Production and the heather moor. *HFRO 5th Report*, 1967-70, 35-44.

GRANT, Sheila A., LAMB, W.I.C., KERR, C.D. and BOLTON, G.R. (1976). The utilisation of blanket bog vegetation by grazing sheep. *J. Appl. Ecol.*, **13**, 857-869.

GRANT, Sheila A. and MILNE, J.A. (1973). Factors affecting the role of heather in grazing systems. In: *Hill pasture improvement and its economic utilisation*. Colloquium Proceedings No. 3 Potassium Institute Ltd., Edinburgh, 1972, 41-46.

GRAY*, E.G. and NICHOLSON, I.A. (1957). Snow mould on upland pasture in North Scotland. *Trans. Bot. Soc. Edinb.*, **37**, 123-128.

GRIFFITHS, J.G. (1967). Behavioural and physiological responses of sheep to climatic exposure. *HFRO 4th Report*, 1964-67, 85-90.

GRIFFITHS, J.G. (1968). Observations on neonatal changes of body temperature in Blackface lambs. *Anim. Prod.*, **10**, 319-323.

GRIFFITHS, J.G. (1970). A note on patterns of behaviour and grazing in hill sheep. *Anim. Prod.*, **12**, 521-524.

GRIFFITHS, J.G. and DONEY, J.M. (1969). Physiological adjustment to repeated wind exposure in sheep. *Anim. Prod.*, **11**, 493-498.

GRIFFITHS, J.G., GUNN, R.G. and DONEY, J.M. (1970). Fertility in Scottish Blackface ewes as influenced by climatic stress. *J. Agric. Sci., Camb.*, **75**, 485-488.

GUNN, R.G. (1962). *A study of the effect of different levels of winter feeding of North and South Country Cheviot ewe hoggs on their growth, development and subsequent performance.* Ph.D. Thesis, University of Edinburgh.

GUNN, R.G. (1963). Effects of treatment during first winter on performance of Cheviot ewes. *Anim. Prod.*, **5**, 244. (summary)

GUNN, R.G. (1964a). Levels of first winter feeding in relation to performance of Cheviot hill ewes. I. Body growth and development during treatment period. *J. Agric. Sci.*, **62**, 99-122.

GUNN, R.G. (1964b). Levels of first winter feeding in relation to performance of Cheviot hill ewes. II. Body growth and development during the summer after treatment, 12-18 months. *J. Agric. Sci.*, **62**, 123-149.

GUNN, R.G. (1965). Levels of first winter feeding in relation to performance of Cheviot hill ewes. III. Tissue and joint development to 12 and 18 months of age. *J. Agric. Sci.*, **64**, 311-321.

GUNN, R.G. (1967a). Levels of first winter feeding in relation to performance of Cheviot hill ewes. IV. Body growth and development from 18 months to maturity. *J. Agric. Sci., Camb.*, **69**, 341-344.

GUNN, R.G. (1967b). Levels of first winter feeding in relation to performance of Cheviot hill ewes. V. Dental development and persistence. *J. Agric. Sci., Camb.*, **69**, 345-348.

GUNN, R.G. (1967c). Life-time performance of the breeding ewe. *HFRO 4th report*, 1964-67, 51-58.

GUNN, R.G. (1967d). A note on hill ewe mortality. *Anim. Prod.*, **9**, 263-264.

GUNN, R.G. (1968a). Levels of first winter feeding in relation to performance of Cheviot hill ewes. VI. Life-time production from the hill. *J. Agric. Sci., Camb.*, **71**, 161-166.

GUNN, R.G. (1968b). A note on difficult birth in Scottish hill flocks. *Anim. Prod.*, **10**, 213-215.

GUNN, R.G. (1968c). Relationships between environment, management and life-time production of hill sheep. In: *Hill-land Productivity.* Br. Grassld Soc. Occ. Symp., **4**, 177-183.

GUNN, R.G. (1969a). The effects of calcium and phosphorus supplementation on the performance of Scottish Blackface hill ewes, with particular reference to the premature loss of permanent incisor teeth. *J. Agric. Sci., Camb.*, **72**, 371-378.

GUNN, R.G. (1969b). A note on seasonal and age changes in the calcium, phosphorus and magnesium content of the blood of Scottish Blackface ewes, as influenced by calcium and phosphorus supplementation. *J. Agric. Sci., Camb.*, **73**, 159-160.

GUNN, R.G. (1969c). A note on the effect of 'broken mouth' on the performance of Scottish Blackface hill ewes. *Anim. Prod.*, **12**, 517-520.

Gunn, R.G. (1972). Growth and lifetime production of Scottish Blackface hill ewes in relation to the level of feeding during rearing. *Anim. Prod.*, **14**, 343-349.

Gunn, R.G. (1975). The effect of extensive grazing systems on the productive potential of hill sheep in Britain. *Proc. III World Conf. Anim. Prod.*, Melbourne, 1973, ed. by R.F. Reid. Sydney Univ. Press, 251-255.

Gunn, R.G. (1977a). The effects of two nutritional environments from six weeks pre-partum to twelve months of age on lifetime performance and reproductive potential of Scottish blackface ewes in two adult environments. *Anim. Prod.*, **25**, 155-164.

Gunn, R.G. (1977b). Reproductive potential of hill sheep. *Vet. Rec.*, **101**, 235. (Summary)

Gunn, R.G. and Doney, J.M. (1964). The sheep of St. Kilda: lessons from feral populations. *HFRO 3rd Report*, 1961-1964, 117-125.

Gunn, R.G. and Doney, J.M. (1970). Some factors affecting reproductive rate in hill sheep. *HFRO 5th Report*, 1967-70, 45-54.

Gunn, R.G. and Doney, J.M. (1973a). The effects of nutrition and rainfall at mating time on the reproductive performance of ewes. *J. Reprod. Fert.*, suppl. 19, 252-258.

Gunn, R.G. and Doney, J.M. (1973b). A note on the influence of the pattern of live-weight and body condition recovery between weaning and mating on food consumption and reproductive performance of Scottish Blackface ewes. *J. Agric. Sci., Camb.*, **81**, 189-191.

Gunn, R.G. and Doney, J.M. (1974). Reproductive response to variation in body condition at mating in three breeds of hill sheep. *Proc. Br. Soc. Anim. Prod.*, **3**, 96. (Abstract)

Gunn, R.G. and Doney, J.M. (1975). The interaction of nutrition and body condition at mating on ovulation rate and early embryonic mortality in Scottish Blackface ewes. *J. Agric. Sci., Camb.*, **85**, 465-470.

Gunn, R.G. and Doney, J.M. (1979). Fertility in Cheviot ewes. I. The effect of body condition at mating on ovulation rate and early embryo mortality in North and South Country Cheviot ewes. *Anim. Prod.* (In press)

Gunn, R.G., Doney, J.M. and Russel, A.J.F. (1969). Fertility in Scottish Blackface ewes as influenced by nutrition and body condition at mating. *J. Agric. Sci., Camb.*, **73**, 289-294.

Gunn, R.G., Doney, J.M. and Russel, A.J.F. (1972). Embryo mortality in Scottish Blackface ewes as influenced by body condition at mating and by post-mating nutrition. *J. Agric. Sci., Camb.*, **79**, 19-25.

Gunn, R.G., Doney, J.M. and Smith, W.F. (1979a). The effect of time of mating on ovulation rate and potential lambing rate of Greyface ewes. *Anim. Prod.* (In press)

Gunn, R.G., Doney, J.M. and Smith, W.F. (1979b). Fertility in Cheviot ewes. II. The effect of level of pre-mating nutrition on ovulation rate and early embryo mortality in North and South Country Cheviot ewes in moderately good condition at mating. *Anim. Prod.* (In press)

GUNN, R.G., DONEY, J.M. and SMITH, W.F. (1979c). Fertility in Cheviot ewes. III. The effect of level of nutrition before and after mating on ovulation rate and early embryo mortality in South Country Cheviot ewes in moderate condition at mating. *Anim. Prod.* (In press)

GUNN, R.G. and MAXWELL, T.J. (1978). The effects of direction of live-weight change about mating on lamb production in Greyface ewes. *Anim. Prod.*, **26**, 392. (Abstract)

GUNN, R.G. and ROBINSON, J.F. (1963). Lamb mortality in Scottish hill flocks. *Anim. Prod.*, **5**, 67-76.

HALLIDAY*, R., RUSSEL, A.J.F., WILLIAMS*, M.R. and PEART, J.N. (1978). The effects of energy intake during late pregnancy and of genotype on immuno-globulin transfer to calves in suckler herds. *Res. Vet. Sci.*, **24**, 26-31.

HALLIDAY*, R., SYKES*, A.R., SLEE*, J., FIELD*, A.C. and RUSSEL, A.J.F., (1969). Cold exposure of Southdown and Welsh Mountain sheep. 4. Changes in concentrations of free fatty acids, glucose, acetone, protein-bound iodine, protein and antibody in the blood. *Anim. Prod.*, **11**, 479-491.

HAMILTON, W.J. (1975). Farming red deer in the Eastern Grampians. *Blackface J.*, **27**, 37-44.

HAMILTON, W.J. joint author (1976). Provisional notes for guidance of Veterinary Officers and other official staff concerned with the requirements of red deer in controlled farming systems. DAFS. (Restricted)

HAYSTEAD, A. (1979). Measurement of nitrogen-15 by mass spectrometry. In: *Modern instrumental methods in soil analysis*, ed. by K.A. Smith. New York, Marcel Dekker. (In press)

HAYSTEAD, A. (1979). Symbiotic nitrogen fixation. In: *Physiological processes limiting crop productivity*, ed. by C. Johnson. Sevenoaks, Butterworths. (In press)

HAYSTEAD, A., KING, J. and LAMB, W.I.C. (1979). Photosynthesis, respiration and nitrogen fixation in white clover. *Grass and Forage Sci.* (In press)

HAYSTEAD, A., KING, J., LAMB, W.I.C. and MARRIOTT, C. (1979). Growth and carbon economy of nodulated white clover in the presence and absence of combined nitrogen. *Grass and Forage Sci.* (In press)

HAYSTEAD, A. and LOWE, A.G. (1977). Nitrogen fixation by white clover in hill pasture. *J. Br. Grassld Soc.*, **32**, 57-63.

HAYSTEAD, A. and MARRIOTT, Carol A. (1978). The fixation and transfer of nitrogen in a clover grass sward under hill conditions. *Ann. Appl. Biol.*, **88**, 453-57.

HAYSTEAD, A. and MARRIOTT, C. (1979a). Effects of different rates and times of application of 'starter' dressings of nitrogen fertiliser to a surface sown perennial ryegrass/white clover sward on hill peat. *Grass and Forage Sci.* (In press)

HAYSTEAD, A. and MARRIOTT, C. (1979b). The transfer of legume nitrogen to an associated grass species. *Soil Biol. Biochem.*, **11**, 99-104.

HEMINGWAY*, R.G., RITCHIE*, N.S., BROWN*, Nora A. and PEART, J.N. (1965). Effects of grazing management on plasma calcium and magnesium concentrations of ewes in early lactation. *J. Agric. Sci., Camb.*, **64**, 109-113.

HODGSON, J. (1976a). The influence of grazing pressure and stocking rate on herbage intake and animal performance. In: *Pasture utilisation by the grazing animal.* Br. Grassld Soc. Occ. Symp., **8**, 93-103.

HODGSON, J. (1976b). Grazing trial design. In: *Improvement and utilisation of grazing resources in Mediterranean-type climates.* Procs. Int. Seminar, Badajoz, Spain, 7-13 May, 1975. Badajoz. Nat. Inst. Agra. Res. (with *INIA/UNDP/FAO*), 180-184.

HODGSON, J. (1977). Factors limiting herbage intake by the grazing animal. *Proc. Int. Meeting on Animal Production from Temperate Grassland*, Dublin, June, 70-75.

HODGSON, J. (1979a). Nomenclature and definitions in grazing studies. *Grass and Forage Sci.*, **34**, 11-18.

HODGSON, J. (1979b). Utilisation of grassland. British Council Course 729: *Management and Diseases of Sheep*, Edinburgh, March 1978. (In press)

HODGSON, J. and MILNE, J.A. (1978). The influence of weight of herbage per unit area and per animal upon the grazing behaviour of sheep. *7th General Meeting, European Grassld Fed.*, Ghent, 4.31-4.38.

HODGSON, J., PEART, J.N., RUSSEL, A.J.F., WHITELAW, A. and MACDONALD, A.J. (1978). The influence of plane of nutrition in early lactation on the performance of spring calving suckler cows and their calves. 2. Grazing period. *Anim. Prod.*, **26**, 381. (Abstract)

HODGSON, J. and WADE*, M.H. (1979). Grazing management and herbage production. In: *Grazing: sward production and livestock output,* papers presented at the Br. Grassld Soc. Winter Meeting, 1978, 1.1-1.12.

HOLDING*, A.J. and KING, J. (1963). The effectiveness of indigenous populations of *Rhizobium trifolii* in relation to soil factors. *Pl. Soil*, **18**, 191-198.

HUGHES, Roy and NICHOLSON, I.A. (1961). Maintenance of reclaimed land. *Scott. Agric.*, **40**, 152-156.

HUGHES, Roy and NICHOLSON, I.A. (1961/2). Comparisons of grass varieties for surface seeding upland pasture types. *J. Br. Grassld Soc.*

    I.   *Molinia* and *Nardus* pastures, **16**, 210-221.
    II. Deep peat, **16**, 314-322.
    III. *Calluna* heath   } **17**, 74-84.
    VI. General discussion }

HUNTER, R.F. (1958a). The direction and problems of research in hill pasture improvement. *J. Br. Grassld Soc.*, **13**, 121-125.

HUNTER, R.F. (1958b). Hill land improvement. *Adv. Sci., Lond.*, **59**, 194-196.

HUNTER, R.F. (1960a). Aims and methods in grazing-behaviour studies on hill pastures. *Proc. 8th Int. Grasssld Cong.*, 454-457.

HUNTER, R.F. (1960b). Conservation and grazing. *Inst. Biol. J.*, **7**, 22-26.

HUNTER, R.F. (1961). Some problems in conducting experiments in hill pasture research. *HFRO 2nd Report*, 1958-61. 77-82.

HUNTER, R.F. (1962). Hill sheep and their pasture: a study of sheep-grazing in South-East Scotland. *J. Ecol.*, **50**, 651-680.

HUNTER, R.F. (1964a). Home range behaviour in hill sheep. In: *Grazing in terrestrial and marine environments,* a symposium of the British Ecological Society,

1962, ed. by D.J. Crisp. Oxford, Blackwells, 155-171.

HUNTER, R.F. (1964b). Land use in the Scottish highlands. VIII. Social behaviour and grazing management in hill sheep. *Adv. Sci., Lond.*, **21**, 169-173.

HUNTER, R.F. and DAVIES, G.E. (1963). The effect of method of rearing on the social behaviour of Scottish blackface hoggets. *Anim. Prod.*, **5**, 183-194.

HUNTER, R.F. and EADIE, J. (1962). Botanical research at the Hill Farming Research Organisation. *Proc. European Conf. for Forage Production on Natural Grassland in Mountain Regions*, 1-7.

HUNTER, R.F. and GRANT, Sheila A. (1961). The estimation of 'green dry matter' in a herbage sample by methanol-soluble pigments. *J. Br. Grassld Soc.*, **16**, 43-45.

HUNTER, R.F. and GRANT, Sheila A. (1971). The effect of altitude on grass growth in East Scotland. *J. Appl. Ecol.*, **8**, 1-19.

HUNTER, R.F. and MILNER, C. (1963). The behaviour of individual, related and groups of South Country Cheviot hill sheep. *Anim. Behav.*, **11**, 507-513.

HYVARINEN*, H., KAY*, R.N.B. and HAMILTON, W.J. (1977). Variation in the weight, specific gravity and composition of the antlers of red deer. *(Cervus elaphus L.) Br. J. Nutr.*, **38**, 301-311.

JAKUBEC*, V., LOUDA*, F., DONEY, J.M. and KRIZEC*, J. (1977). The quantitative and qualitative milk production of the improved Valachian breed and its crosses with Finnsheep. *28th Ann. Meeting EAAP*, Brussels, 1977.

JENKINS*, R., HURLEY*, P.W. and SHORROCKS, V.M. (1966). Plant mineral analysis by X-ray fluorescence spectrometry. *Analyst, Lond.*, **91**, 395-397.

JONES, D. Neil, (1958a). Performance of Blackface sheep at Glensaugh. *Trans. R. Highl. Agric. Soc. Scott.*, 6th series, **3**, 87-89.

JONES, D. Neil, (1958b). A survey of management systems and the incidence of 'broken mouths' in Blackface sheep in the North-East of Scotland. *HFRO 1st Report*, 1954-58, 57-71.

KING, J. (1955). *The ecology of certain communities in the Cheviots with particular reference to Nardus stricta.* Ph.D. Thesis, University of Edinburgh.

KING, J. (1960). Observations on the seedling establishment and growth of *Nardus stricta* in burned Callunetum. *J. Ecol.*, **48**, 667-677.

KING, J. (1961a). Ecotypic differentiation in *Trifolium repens. Pl. Soil*, **18**, 221-224.

KING, J. (1961b). Range research in the Pacific North West of the United States. *HFRO 2nd report*, 1958-61, 92-96.

KING, J. (1962). The *Festuca/Agrostis* grassland complex in South-East Scotland. *J. Ecol.*, **50**, 321-355.

KING, J. (1971). Competition between established and newly sown grass species. *J. Br. Grassld Soc.*, **26**, 221-229.

KING, J. (1977). Hill and upland pasture. In: *Rural land management and agricultural change: a handbook of applied ecology*, ed. by J. Davidson and R. Lloyd, London, Wiley, 95-119.

KING, J. and DAVIES, G.E. (1963). The effect of dalapon on the species of hill grassland. *J. Br. Grassld Soc.*, **18**, 52-55.

King, J., Grant, Sheila A. and Rogers, J.A. (1967). Hill pastures. *HFRO 4th Report*, 1964-67, 22-32.

King, J., Lamb, W.I.C. and McGregor, M.T. (1978). Effect of partial and complete defoliation on regrowth of white clover plants. *J. Br. Grassld Soc.*, **33**, 49-55.

King, J., Lamb, W.I.C. and McGregor, M.T. (1979). Regrowth of ryegrass swards subject to different cutting regimes and stocking rates. *Grass and Forage Sci.* (In press)

King, J. and Nicholson, I.A. (1964). The grasslands of the forest zone. In: *The vegetation of Scotland*, ed. by C.H. Burnett. Edinburgh, Oliver and Boyd, 168-206.

Krizek*, J., Louda*, F., Jakubec*, V. and Doney, J.M. (1978). [Analysis of milk production in the improved Valachian breed, prolific breeds and their crossbreds.] *Zivocisna Vyroba*, **23**, 461-470. (Czech)

Lamb, C.S. and Eadie, J. (1979). The effect of barley supplements on the voluntary intake and digestion of low quality roughages by sheep. *J. Agric. Sci., Camb.*, **92**, 235-241.

Le Du, Y.L.P., Peart, J.N. and Macdonald, A.J. (1979). Evaluation of some aspects of a machine milking technique for estimating the milk production of beef cows. *Livest. Prod. Sci.*, **6**. (In press)

Louda*, F. and Doney, J.M. (1976). Persistency of lactation in the improved Valachian breed of sheep. *J. Agric. Sci., Camb.*, **87**, 455-457.

Louda*, F., Doney, J.M., Jakubec*, V., Picha*, J. and Krizek*, J. (1975). [Factors influencing the reproductive performance of ewes with special reference to semen quality of the ram and nutrition and body condition of the ewe.] *Proceedings of International Specialist Conference*, Prague, 1975, Sekce Zootech., 459-467. (Czech)

Louda*, F., Doney, J.M., Smerha*, J. and Skrivan*, M. (1977). The dynamics of sperm production in Merino rams in relation to season. *28th Ann. Meeting, EAAP*, Brussels, 1977.

Louda*, F., Jakubec*, V., Krizek*, J. and Doney, J.M. (1978). Possibilities of influencing ewe fertility by controlled nutrition. *Scient. Agric. Bohem.*, **10**, 131-136.

Louda*, F., Krizek*, J. and Doney, J.M. (1979). [Milk production and lamb growth in the Asch Merino.] *Zivocisna Vyroba*. (In press) (Czech.)

McClelland*, T.H. and Russel, A.J.F. (1971). Content and distribution of body fat in Finnish Landrace and Scottish Blackface wether lambs. *Anim. Prod.*, **13**, 395A. (Abstract)

McClelland*, T.H. and Russel, A.J.F. (1972). The distribution of body fat in Scottish Blackface and Finnish Landrace lambs. *Anim. Prod.*, **15**, 301-306.

McClelland*, T.H., Russel, A.J.F. and Jackson*, T.H. (1973). Lamb growth, efficiency of food utilisation and body fat at four stages of maturity in four breeds of different mature body weight. *Proc. Br. Soc. Anim. Prod.*, **2**, 83. (Abstract)

Macleod, A.C. (1955). Heather in the seasonal dietary of sheep. *Proc. Br. Soc. Anim. Prod.*, 13-17.

Macrae, J.C. (1974). The use of intestinal markers to measure digestive function in ruminants. *Proc. Nutr. Soc.*, **33**, 147-155.

MACRAE, J.C. (1975). The use of re-entrant cannulae to partition digestive function within the gastro-intestinal tract of ruminants. In: *Digestion and metabolism in the ruminant,* ed. by I.W. McDonald and A.C.I. Warner. Proceedings of the IVth International Symposium on Ruminant Physiology, Sydney, 1975, 261-276.

MACRAE, J.C. (1976). Utilisation of the protein of green forage by ruminants at pasture. In: *From plant to animal protein* — Rural Science Reviews II, ed. by T.M. Sutherland, J.R. McWilliams and R.A. Leng, Armidale, Australia, University of New England Publishing Unit, 1976, 93-98.

MACRAE, J.C. (1979). The effects of processing on nutrient content of feeds: freezing. In: *CRC Handbook of Nutrition and Food.* (In press)

MACRAE, J.C., CAMPBELL, D.R. and EADIE, J. (1975). Changes in biochemical composition of herbage upon freezing and thawing. *J. Agric. Sci., Camb.,* **84,** 125-132.

MACRAE, J.C. and EVANS, C.C. (1974). The use of inert ruthenium-phenanthroline as a digesta particulate marker in sheep. *Proc. Nutr. Soc.,* 33, 10A. (Abstract)

MACRAE, J.C., MILNE, J.A., WILSON, S. and SPENCE, A.M. (1979). Nitrogen digestion in sheep given poor-quality indigenous hill herbage. *Br. J. Nutr.* (In press)

MACRAE, J.C. and WILSON, S. (1977). The effects of various forms of gastro-intestinal cannulation on digestive parameters in sheep. *Br. J. Nutr.,* **38,** 65-71.

MACRAE, J.C. and WILSON, S. (1978). Problems associated with scintillation counting of $NaH_{14}CO_3$ and gel suspension counting of $Ba_{14}CO_3$. *Int. J. Appl. Radiat. Isot.,* **29,** 191-195.

MACRAE, J.C., WILSON, S. and MILNE, J.A. (1976). Carbon metabolism in sheep fed on poor quality hill herbage. *Proc. Nutr. Soc.,* **35,** 103A. (Abstract)

MACRAE, J.C., WILSON, S. and MILNE, J.A. (1978). Microbial and host-animal components of energy metabolism in hill sheep. *Proc. Nutr. Soc.,* **37,** 16A. (Abstract)

MACRAE, J.C., WILSON, S., MILNE, J.A. and SPENCE, A.M. (1977). Urea-nitrogen recycling in sheep given low quality hill herbages. *Proc. Nutr. Soc.,* **36,** 77A. (Abstract)

MARTIN*, A.K., MILNE, J.A. and MOBERLY, Patricia (1975). Urinary quinol and orcinol outputs as indicies of voluntary intake of heather (*Calluna Vulgaris L. Hull*) by sheep. *Proc. Nutr. Soc.,* **34,** 70A-71A. (Abstract)

MAXWELL, T.J. (1977a). Perinatal losses in sheep. Cost benefit implications: effect on hill return. *Proc. ESCA Symposium on Perinatal Losses in Sheep.* Stirling, 5-6 February, 56-58.

MAXWELL, T.J. (1977b). Utilisation of hill land. *Vet. Rec.,* **101,** 207. (Summary)

MAXWELL, T.J: (1978). Upland and lowland sheep: production problems. *Vet. Rec.,* **103,** 177. (Abstract)

MAXWELL, T.J. (1979). Econometric techniques for the assessment of sheep systems. Brit. Council Course 729: *Management and Diseases of Sheep,* Edinburgh, March 1978. (In press)

MAXWELL, T.J., DONEY, J.M., MILNE, J.A., PEART, J.N., RUSSEL, A.J.F., SIBBALD,

A.R. and MacDonald, D. (1979). The effect of rearing type and pre-partum nutrition on the intake and performance of lactating Greyface ewes at pasture. *J. Agric. Sci., Camb.,* **92,** 165-174.

Maxwell, T.J., Eadie, J. and Sibbald, A.R. (1973a). Economic appraisal of investments in hill sheep production. *HFRO 6th Report,* 1971-73, 24-42.

Maxwell, T.J., Eadie, J. and Sibbald, A.R. (1973b). Methods of economic appraisal of hill sheep production systems. In: *Hill pasture improvement and its economic utilisation.* Colloquium Proceedings No. 3 Potassium Institute Ltd., Edinburgh, 1972, 103-113.

Maxwell, T.J., Eadie, J. and Sibbald, A.R. (1979). Some economic implications in the application of new techniques to hill sheep farming in the United Kingdom. *Int. Hill Land Symposium,* Morgantown, West Virginia, 3-9 October, 1976. (In press)

Maxwell, T.J., Sibbald, A.R. and Eadie, J. (1979). The integration of forestry and agriculture — a model. *Agric. Systems.* (In press)

Milne, J.A. (1974). The effect of season and age of stand on the nutritive value of heather *(Calluna vulgaris)* to sheep. *J. Agric. Sci., Camb.,* **83,** 281-289.

Milne, J.A. (1977a). A comparison of methods of predicting the *in vivo* digestibility of heather by sheep. *J. Br. Grassld Soc.,* **32,** 141-147.

Milne, J.A. (1977b). Diet selection by the grazing sheep. *Br. Vet. J.,* **133,** 96. (Abstract)

Milne, J.A. and Bagley, L. (1976). The nutritive value of diets containing different proportions of grass and heather *(Calluna vulgaris L. Hull)* to sheep. *J. Agric. Sci., Camb.,* **87,** 599-604.

Milne, J.A., Bagley, L. and Grant, Sheila A. (1979). Effect of season and level of grazing on the utilisation of heather by sheep. 2. Diet selection and intake by sheep. *Grass and Forage Sci.,* **34,** 45-53.

Milne, J.A., Christie, A. and Russel, A.J.F. (1979). The effects of nitrogen and energy supplementation on the voluntary intake and digestion of heather by sheep. *J. Agric. Sci., Camb.,* (In press)

Milne, J.A. and Grant, Sheila A. (1978). Better use of heather hills for sheep production. *HFRO 7th Report,* 1974-77, 41-48.

Milne, J.A., Macrae, J.C., Spence, Angela M. and Wilson, S. (1976). Intake and digestion of hill land vegetation by the red deer and the sheep. *Nature, Lond.,* **263,** 763.

Milne, J.A., Macrae, J.C., Spence, A.M. and Wilson, S. (1978). A comparison of the voluntary intake and digestion of a range of forages at different times of the year by the sheep and the red deer *(Cervus elaphus)*. *Br. J. Nutr.,* **40,** 347-357.

Milne, J.A., Maxwell, T.J. and Souter, W. (1979). Effect of level of concentrate feeding and amount of herbage on the intake and performance of ewes with twin lambs at pasture in early lactation. *Anim. Prod.* (Abstract) (In press)

Milner, C. (1963). *The diet and behaviour of hill sheep.* Ph.D. Thesis, University of Edinburgh.

Munro, Joan (1961). Shelter for livestock: a survey in two hill areas in Scotland. *HFRO 2nd Report,* 1958-61, 83-91.

MUNRO, Joan (1962a). A study of the milk yield of three strains of Scottish Black-face ewes in two environments. *Anim. Prod., 4,* 203-213.

MUNRO, Joan (1962b). The use of natural shelter by hill sheep. *Anim. Prod., 4,* 343-349.

NEWBOULD, P. (1973). Plants to improve hill pastures. *HFRO 6th Report,* 1971-73, 74-85.

NEWBOULD, P. (1974/5). Improvement of hill pastures for agriculture.
A review.   Pt. 1 *J. Br. Grassld Soc., 29,* 241-248.
Pt. 2 *J. Br. Grassld Soc., 30,* 41- 44.

NEWBOULD, P. (1977). Principles of nutrient cycling, Chapter 2 in: *The cycling of mineral nutrients in agricultural ecosystems. Agro-ecosystems, 4,* 3-6.

NEWBOULD, P. (1978). The work of the Hill Farming Research Organisation. *Proc. N.Z. Grassld Assoc., 9.*

NEWBOULD, P. (1979). Integrating agriculture and forestry — a British view. In: *Integrating agriculture and forestry,* Proc. of Workshop, Bunbury, Western Australia, Sept. 1977, CSIRO, 141-159.

NEWBOULD, P. (1979). Techniques for hill land improvement used in the United Kingdom. *International Hill Land Symposium,* Morgantown, West Virginia, 3-9 October, 1976. (In press)

NEWBOULD, P. and FLOATE, M.J.S. (1977). Nutrient cycling in agro-ecosystems in the U.K., Chapter 6 in: *The cycling of mineral nutrients in agricultural ecosystems. Agro-Ecosystems, 4,* 33-70.

NEWBOULD, P. and FLOATE, M.J.S. (1979). Problems of hill and upland soils. *Proc. Welsh Soils Discussion Group.* (In press)

NEWBOULD, P., FRISSEL*, M. and FLOATE, M.J.S. (1977). Nutrient cycling summary and conclusions. Chapter 8 in: *The cycling of mineral nutrients in agricultural ecosystems. Agro-Ecosystems, 4,* 313-315.

NEWBOULD, P. and HAYSTEAD, A. (1978). *Trifolium repens* (white clover): its role, establishment and maintenance in hill pastures. *HFRO 7th Report,* 1974-77, 49-68.

NICHOLSON, I.A. (1957). The effect of stage of maturity on the yield and chemical composition of oats for haymaking. *J. Agric. Sci., 49,* 129-140.

NICHOLSON, I.A. (1959a). Aircraft in agriculture. *Scott. Agric., 39,* 127-132.

NICHOLSON, I.A. (1959b). Hill pasture improvement and the significance of aircraft. *Proc. 1st Int. Agric. Aviation Conf.,* 79-86.

NICHOLSON, I.A. (1960a). Aerial farming. *Fmg Rev., 15,* 6-10.

NICHOLSON, I.A. (1960b). Economic aspects of bracken. *5th Br. Weed Control Conf. Proc., 1,* 179-186.

NICHOLSON, I.A. (1964a). Land use in the Scottish highlands. IV. The influence of management practices on the present day vegetational pattern and developmental trends. *Adv. Sci., Lond., 21,* 158-163.

NICHOLSON, I.A. (1964b). Some range and watershed problems in the United States. *HFRO 3rd Report,* 1961-64, 103-116.

NICHOLSON, I.A. (1967). The grazing animal in vegetational control. *HFRO 4th Report,* 1964-67, 46-50.

NICHOLSON, I.A. (1968). Some concepts in the use and improvement of range grazings in the British uplands. In: *Hill-land Productivity*. Br. Grassld Soc. Occ. Symp., **4**, 58-61.

NICHOLSON, I.A., CURRIE, D.C., PATERSON, I.S. and McCREATH*, J.B. (1968). Hill grazing management and increased production. *Scott. Agric.*, **47**, 123-131.

NICHOLSON, I.A. and HUGHES, Roy (1963). A method for the characterisation of range-type vegetation. *J. Range Mgmt*, **16**, 293-299.

NICHOLSON, I.A., PATERSON, I.S. and CURRIE, A. (1970). Vegetational dynamics: selection by sheep and cattle in a *Nardus* pasture. In: *Animal populations in relation to their food resources*, ed. by A. Watson. Br. Ecol. Soc. Symp. No. 10. Oxford, Blackwell.

NICHOLSON, I.A., PATERSON, I.S. and CURRIE, D.C. (1962). Electric fencing for hill land. *Scott. Agric.*, **42**, 73-76.

NICHOLSON, I.A. and ROBERTSON*, R.A. (1958). Some observations on the ecology of an upland grazing in North-East Scotland with special reference to *Callunetum*. (Survey of Glensaugh). *J. Ecol.*, **46**, 239-270.

NICHOLSON, I.A. and SAMFORD*, M.R. (1958). The effect of some treatments in August on the performance of a pasture sward between September and January and the subsequent effects in spring. *HFRO 1st Report*, 1954-58, 86-88.

NICHOLSON, I.A. and SMITH*, R.G.C. (1957). The use of autumn grass for weaned calves. *Proc. Br. Soc. Anim. Prod.*, 33-40.

NOLAN*, J.V. and MACRAE, J.C. (1976). Absorption and recycling of nitrogenous compounds in the digestive tract of sheep. *Proc. Nutr. Soc.*, **35**, 110A. (Abstract)

PEART, J.N. (1958). A survey of Border Cheviot sheep. *HFRO 1st Report*, 1954-58, 72-85.

PEART, J.N. (1963). Increased production from hill pastures. Sourhope trials with cattle and sheep. *Scott. Agric.*, **42**, 147-151.

PEART, J.N. (1967a). The effect of different levels of nutrition during late pregnancy on the subsequent milk production of Blackface ewes and on the growth of their lambs. *J. Agric. Sci.*, **68**, 365-371.

PEART, J.N. (1967b). Lactation and lamb growth. *HFRO 4th Report*, 1964-67, 69-76.

PEART, J.N. (1968a). Lactation studies with Blackface ewes and their lambs. *J. Agric. Sci., Camb.*, **70**, 87-94.

PEART, J.N. (1968b). Some effects of live weight and body condition on the milk production of Blackface ewes. *J. Agric. Sci., Camb.*, **70**, 331-338.

PEART, J.N. (1970a). Factors influencing lactation of hill ewes. *HFRO 5th Report*, 1967-70, 55-69.

PEART, J.N. (1970b). The influence of live weight and body condition on the subsequent milk production of Blackface ewes following a period of under-nourishment in early lactation. *J. Agric. Sci., Camb.*, **75**, 459-469.

PEART, J.N. (1970c). Sheep production in relation to stocking rates and controlled grazing of hill pastures. *Expl Husb.*, **19**, 29-39.

PEART, J.N., DONEY, J.M. and MACDONALD, A.J. (1975). The influence of lamb genotype on the milk production of Blackface ewes. *J. Agric. Sci., Camb.*, **84**, 313-316.

PEART, J.N., DONEY, J.M. and SMITH, W.F. (1979). Lactation pattern in Scottish Blackface and East Friesland × Scottish Blackface cross-bred ewes. *J. Agric. Sci., Camb.*, **92**, 133-138.

PEART, J.N., EDWARDS*, R.A. and DONALDSON*, Elizabeth (1972). The yield and composition of the milk of Finnish Landrace × Blackface ewes. I. Ewes and lambs maintained indoors. *J. Agric. Sci., Camb.*, **79**, 303-313.

PEART, J.N., EDWARDS*, R.A. and DONALDSON*, Elizabeth (1975). The yield and composition of the milk of Finnish Landrace × Blackface ewes. II. Ewes and lambs grazed on pasture. *J. Agric. Sci., Camb.*, **85**, 315-324.

PEART, J.N., RUSSEL, A.J.F., HODGSON, J., WHITELAW, A. and MACDONALD, A.J. (1978). The influence of plane of nutrition in early lactation on the performance of spring-calving suckler cows and their calves. I. Winter period. *Anim. Prod.*, **26**, 380-381. (Abstract)

PEART, J.N. and RYDER*, M.L. (1954). Some observations on different fleece types in Scottish Blackface sheep. *J. Text. Inst. Manchr*, **45**, 1821-1827.

POLLOCK*, D.L., PRESCOTT*, J.H.D., DONEY, J.M. and SMITH*, C. (1976). Group breeding schemes for Blackface sheep. *Blackface J.*, **28**, 13-17.

REID, R.L. (1968). The physio-pathology of undernourishment in pregnant sheep, with particular reference to pregnancy toxaemia. *Adv. Vet. Sci.*, **12**, 163-238.

ROBERTSON*, R.A. and DAVIES, G.E. (1965). Quantities of plant nutrients in heather ecosystems. *J. Appl. Ecol.*, **2**, 211-219.

ROBERTSON*, R.A. and NICHOLSON, I.A. (1961). The response of some hill pasture types to lime and phosphate. *J. Br. Grassld Soc.*, **16**, 117-125.

ROBERTSON*, R.A., NICHOLSON, I.A. and HUGHES, R. (1963). Runoff studies on a peat catchment. *International Peat Congress*, Section 2, Leningrad, 1-8.

ROBINSON, J.F. (1955/56). Sheep husbandry in Iceland. *Scott. Agric.* **35**, 132-135.

ROBINSON, J.F., CURRIE, D.C. and PEART, J.N. (1961). Feeding hill ewes. *Trans. Roy. Highl. Agric. Soc. Scott.*, **6**, 31-46.

ROGERS, J.A. (1977). Preliminary assessment of a 'super-sonic' seed sowing device. *J. Agric. Eng. Res.*, **22**, 97-100.

ROGERS, J.A. and DAVIES, G.E. (1973). The growth and chemical composition of four grass species in relation to soil moisture and aeration factors. *J. Ecol.*, **61**, 455-472.

ROGERS, J.A. and KING, J. (1972a). The distribution and abundance of grassland species in hill pasture in relation to soil aeration and base status. *J. Ecol.*, **60**, 1-18.

ROGERS, J.A. and KING, J. (1972b). Soil moisture and aeration in relation to plant growth — a field investigation. *Proc. N. England Soils Discussion Group*, **7**, 28-31.

ROGERS, J.A., KING, J. and DAVIES, G.E. (1973). The problem of waterlogged soil. *HFRO 6th Report*, 1971-73, 86-94.

RUSSEL, A.J.F. (1967a). A note on goitre in lambs grazing rape (*Brassica napus*). *Anim. Prod.*, **9**, 131-133.

RUSSEL, A.J.F. (1967b). Nutrition of the pregnant ewe. *HFRO 4th Report, 1964-67*, 59-68.

RUSSEL, A.J.F. (1968). *Studies in physiological undernourishment in sheep.* Ph.D. Thesis, University of Edinburgh.

RUSSEL, A.J.F. (1969). Nutrition of the pregnant hill ewe. *Ir. Grassld and Anim. Prod. J.*, **4**, 80-88.

RUSSEL, A.J.F. (1971). Relationships between energy intake and productivity in hill sheep. *Proc. Nutr. Soc.*, **30**, 197-204.

RUSSEL, A.J.F. (1973). Recent developments in sheep and beef cattle production in N.Z. *Proc. Br. Soc. Anim. Prod.*, **2**, 95-103.

RUSSEL, A.J.F. (1978a). The relative contributions of nutrition and genetics to improvements in the efficiency of sheep production. *Agric. Prog.*, **58**, 92-97.

RUSSEL, A.J.F. (1978b). The use of measurement of energy status in pregnant ewes. In: *The use of blood metabolites in animal production.* BSAP Occ. Publ. **1**, 31-40.

RUSSEL, A.J.F. (1979). Nutrition of the pregnant ewe. British Council Course 729: *Management and diseases of sheep*, Edinburgh, March, 1978. (In press)

RUSSEL, A.J.F. and BARTON*, R.A. (1967). Bone-muscle relationships in lamb and mutton carcases. *J. Agric. Sci.*, **68**, 187-190.

RUSSEL, A.J.F. and DONEY, J.M. (1969a). Observations on the use of plasma free fatty acid concentrations in the determination of maintenance requirements of sheep. *J. Agric. Sci., Camb.*, **72**, 59-63.

RUSSEL, A.J.F. and DONEY, J.M. (1969b). Measurement of the cost of climatic exposure of sheep. *Third Symp. on Shelter Research, Cambridge.* London, MAFF Land Improvement Division, 103-112.

RUSSEL, A.J.F. and DONEY, J.M. (1969c). The use of plasma free fatty acid concentrations in the determination of energy requirements of sheep. *VIIIth International Nutrition Congress, Prague.*

RUSSEL, A.J.F. and DONEY, J.M. (1974). Production from hill sheep. *Norgrass*, **14**, 8-16.

RUSSEL, A.J.F., DONEY, J.M. and EADIE, J. (1978). Current knowledge and unresolved problems in the nutrition of hill sheep. *HFRO 7th Report, 1974-77*, 29-40.

RUSSEL, A.J.F., DONEY, J.M. and GUNN, R.G. (1969). Subjective assessment of body fat in live sheep. *J. Agric. Sci., Camb.*, **72**, 451-454.

RUSSEL, A.J.F., DONEY, J.M. and GUNN, R.G. (1971). The distribution of chemical fat in the bodies of Scottish Blackface ewes. *Anim. Prod.*, **13**, 503-509.

RUSSEL, A.J.F., DONEY, J.M. and MAXWELL, T.J. (1976). The effect on wool production of changes in management designed to increase output of lamb from hill land in the United Kingdom. *Livest. Prod. Sci.*, **3**, 173-182.

RUSSEL, A.J.F., DONEY, J.M. and REID, R.L. (1967a). The use of biochemical parameters in controlling nutritional state in pregnant ewes, and the effect of undernourishment during pregnancy on lamb birth-weight. *J. Agric. Sci.*, **68**, 351-358.

RUSSEL, A.J.F., DONEY, J.M. and REID, R.L. (1967b). Energy requirements of the pregnant ewe. *J. Agric. Sci.*, **68**, 359-363.

Russel, A.J.F. and Eadie, J. (1968). Nutrition of the hill ewe. In: *Hill-land Productivity*. Br. Grassld Soc. Occ. Symp., **4**, 185-191.

Russel, A.J.F. and Foot, Janet Z. (1971). The use of circulating progesterone concentrations in the identification of twin-bearing ewes. *Anim. Prod.*, **13**, 377A. (Abstract)

Russel, A.J.F. and Foot, Janet Z. (1973). The effect of level of nutrition at two stages of pregnancy on the performance of primiparous ewes. *Proc. Nutr. Soc.*, **32**, 27A. (Abstract)

Russel, A.J.F., Gunn, R.G. and Doney, J.M. (1968). Components of weight loss in pregnant hill ewes during winter. *Anim. Prod.*, **10**, 43-51.

Russel, A.J.F., Gunn, R.G., Skedd, E. and Doney, J.M. (1968). Relationships between chemical and physical compositions of Scottish Blackface ewes. *Anim. Prod.*, **10**, 53-58.

Russel, A.J.F., Macdonald, A.J., Kerr, C.D. and Rudd, Brenda (1976). Changes in live weight and body condition of rams of three breeds throughout the year. *Anim. Prod.*, **23**, 73-80.

Russel, A.J.F., Maxwell, T.J. and Foot, Janet Z. (1973). Nutrition of the hill ewe during late pregnancy. *HFRO 6th Report*, 1971-73, 43-56.

Russel, A.J.F., Maxwell, T.J., Sibbald, A.R. and MacDonald, D. (1977). Relationship between energy intake, nutritional state and lamb birth weight in Greyface ewes. *J. Agric. Sci., Camb.*, **89**, 667-673.

Russel, A.J.F., Peart, J.N., Eadie, J., Macdonald, A.J. and White, I.R. (1979). The effect of energy intake during late pregnancy on the production from two genotypes of suckler cows. *Anim. Prod.* (In press)

Russel, A.J.F., Whitelaw, A., Moberly, Patricia and Fawcett, A.R. (1975). An investigation into the diagnosis and treatment of cobalt deficiency in lambs. *Vet. Rec.*, **96**, 194-198.

Sibbald, A.R., Maxwell, T.J. and Eadie, J. (1979). A conceptual approach to the modelling of herbage intake by sheep. *Agric. Syst.*, **4**, 119-134.

Singer*, M., Holding*, A.J. and King, J. (1964). The response of *Trifolium repens* to inocula containing varying proportions of effective and ineffective Rhizobia. *Trans. 8th Int. Congr. Soil Sci.*, Bucharest, **3**, 1021-1025.

Sykes*, A.R., Field*, A.C. and Gunn, R.G. (1974a). Effects of age and state of incisor dentition on body composition and lamb production of sheep grazing hill pastures. *J. Agric. Sci., Camb.*, **83**, 135-143.

Sykes*, A.R., Field*, A.C. and Gunn, R.G. (1974b). Effects of age and state of incisor dentition on the chemical composition of the skeleton of sheep grazing hill pastures. *J. Agric. Sci., Camb.*, **83**, 145-150.

Sykes*, A.R. and Russel, A.J.F. (1979). Seasonal variation in plasma protein and urea nitrogen concentrations in hill sheep. *Res. Vet. Sci.* (In press)

Tait, Alison, Milne, J.A. and Russel, A.J.F. (1975). Studies on the voluntary intake of heather (*Calluna vulgaris*) by sheep. *Proc. Br. Soc. Anim. Prod.*, **4**, 98-99. (Abstract)

Vernon, D.M. (1978). *The selection of strains of Rhizobium trifolii for inoculation of Trifolium repens for hill pasture*. M. Phil. Thesis, University of Edinburgh.

Vine, D.A. (1977). *Development of a pasture model for grazing studies.* Ph.D. Thesis, University of Edinburgh.

Wannop, A.R. (1954/55). The Hill Farming Research Organisation. *Scott. Agric.,* **34,** 177-180.

Wannop, A.R. (1955a). Animal production from grass. *Proc. Br. Soc. Anim. Prod.,* 115-119.

Wannop, A.R. (1955b). The Hill Farming Research Organisation. *J. Blackface Sheep Breed. Assoc.,* **8,** 17-19.

Wannop, A.R. (1958a). Hill Farming in New Zealand. *J. Black. Sheep Breed. Assoc.,* **11,** 13-17.

Wannop, A.R. (1958b). A review of hill farm problems. *Adv. Sci., Lond.,* **59,** 189-194.

Wannop, A.R. (1958c). Sheep Farming in New Zealand. *Proc. Br. Soc. Anim. Prod.,* 101-103.

Wannop, A.R. (1959). Science and the hill farmer. *Agriculture, Lond.,* **66,** 1-5.

Wannop, A.R. (1960). Hill farming research. *Scott. Agric.,* **40,** 19-21.

Watt*, J.A., Foster, W.N.M. and Cameron, A.E. (1968). Benzathine penicillin as a prophylactic in tick pyaemia. *Vet. Rec.,* **83,** 507-508.

Whitelaw, A. (1976). Survey of perinatal losses associated with intensive hill sheep farming. *Veterinary Annual,* **16,** 60-65.

Whitelaw, A. (1977). Control of liver fluke in hill sheep. *Blackface J.,* **29,** 41.

Whitelaw, A. (1977). Perinatal losses and intensive hill sheep farming. *Proc. ESCA Symp. on Perinatal Losses in Sheep,* Stirling, 5-6 Feb., 1975, 47-49.

Whitelaw, A. (1978). Copper toxicity in sheep. *Vet. Rec.,* **103,** 453. (Letter)

Whitelaw, A. (1979). Cobalt deficiency — aspects of treatment and prevention. *ADAS Trace element seminar,* Newcastle, March 1979. (In press)

Whitelaw, A., Armstrong, R.H., Evans, C.C. and Fawcett, A.R. (1977). An investigation into copper deficiency in young lambs on an improved hill pasture. *Vet. Rec.,* **101,** 229-230.

Whitelaw, A., Armstrong, R.H., Evans, C.C. and Fawcett, A.R. (1979). A study of the effects of copper deficiency in Scottish Blackface lambs on improved hill pasture. *Vet. Rec.,* **104,** 455-460.

Whitelaw, A. and Evans, C.C. (1979). Further investigations into copper deficiency in young lambs grazing improved pastures. Soc. Chemical Industry Agric. Group. *Symposium on trace elements in soils, plants and animals,* London, 1979. (In press)

Whitelaw, A. and Fawcett, A.R. (1976). Anaesthesia in red deer. *Vet. Rec.,* **99,** 424. (Letter)

Whitelaw, A. and Fawcett, A.R. (1977). A study of strategic dosing programme against ovine fascioliasis on a hill farm. *Vet. Rec.,* **100,** 443-447.

Whitelaw, A. and Russel, A.J.F. (1979a). Investigations into the prophylaxis of cobalt deficiency in sheep. *Vet. Rec.,* **104,** 8-11.

Whitelaw, A. and Russel, A.J.F. (1979b). Cobalt deficiency — aspects of prevention. *COSAC Conference on Trace Elements,* Aberdeen. (In press)

WHITELAW, A. and WATCHORN, P. (1975). An investigation into dystocia in a South Country Cheviot flock. *Vet. Rec.*, **97**, 489-492.

WILLIAMS*, C.H., NEWBOULD, P. and FLOATE, M.J.S. (1977). Agro-ecosystems in Australia. Chapter 6 in: *The cycling of mineral nutrients in agricultural ecosystems. Agro-Ecosystems*, **4**, 179-182.

WILSON, S. and MACRAE, J.C. (1977). An investigation into the loss of activity of ^{14}C-labelled volatile fatty acid (VFA) tracers used to measure VFA production rates in the rumen of sheep. *J. Agric. Sci., Camb.*, **88**, 245-246.

WILSON, S., MACRAE, J.C. and BUTTERY*, P.J. (1979). The effects of implanting jugular catheters on plasma glucose concentrations in wethers fed two extremes of diet. *Res. Vet. Sci.*, **26**, 256-258.

Printed by D. & J. Croal Ltd., 20 Market Street, Haddington